Statistics for Data Science and Analytics

Statistics for Data Science and Analytics

Peter C. Bruce
Founder, Institute for Statistics Education at Statistics.com

Peter Gedeck
Senior Data Scientist, Collaborative Drug Discovery

Janet Dobbins
Chair, Data Community DC

WILEY

Published by John Wiley & Sons, Inc., Hoboken, New Jersey.
Published simultaneously in Canada.

Library of Congress Cataloging-in-Publication Data Applied for:

Hardback ISBN: 9781394253807

Cover Design: Wiley
Cover Image: © gremlin/Getty Images

Set in 9.5/12.5pt STIXTwoText by Straive, Chennai, India

To my wife, Liz, whose editorial help and judgment is impeccable—PB

To my parents, Helga and Erhard Gedeck, and my sister, Heike, who have always been supportive of my education—PG

To my friend Peter, whose guidance on my data science journey has been invaluable, and to my husband Peter, whose encouragement adds joy to every step—JD

Contents

About the Authors

Peter C. Bruce is the Founder of the Institute for Statistics Education at Statistics.com. Founded in 2002, Statistics.com was the first educational institution devoted solely to online education in statistics and data science.

Peter Gedeck is a senior data scientist at Collaborative Drug Discovery. He specializes in the development of cloud-based software for managing data in the drug discovery process. In addition, he teaches data science at the University of Virginia.

Janet Dobbins is a leading voice in the Washington, DC, data science community. She is Chair of the Board of Directors for Data Community DC (DC2) and co-organizes the popular Data Science DC meetups. She previously worked at the Institute for Statistics Education at Statistics.com.

Acknowledgments

Julian Simon, an early resampling pioneer, first kindled Peter C. Bruce's interest in statistics with his permutation and bootstrap approach to statistics, his Resampling Stats software (first released in the late 1970s), and his statistics text on the same subject. Robert Hayden, who co-authored early versions of parts of the text you are reading, was instrumental in getting this project launched.

Michelle Everson has taught many sessions of introductory statistics using versions of this book and has been vigilant in pointing out ambiguities and omissions. Her active participation in the statistics education community has been an asset as we have strived to improve and perfect this text. Meena Badade also teaches using this text and has also been very helpful in bringing to our attention errors and points requiring clarification. Diane Murphy reviewed the latest version of the book with care and contributed many useful corrections and suggestions.

Many students at the Institute for Statistics Education at Statistics.com have helped clarify confusing points and refine this book over the years.

We also thank our editor at Wiley, Brett Kurzman, who shepherded this book through the acceptance and launch process quickly and smoothly. Nandhini Karuppiah, the Managing Editor, helped guide us through the production process, and Govind Nagaraj managed the copyediting.

About the Companion Website

This book is accompanied by a companion website:

www.wiley.com/go/Wiley_Statistics_for_Data

We are happy that you have chosen our book for your course. For instructors that adopt the book, we provide these supplemental materials:

- Short answers and exercises in the text.
- Datasets and Python examples
- Videos mentioned in the test
- Link to GitHub repository and Jupyter notebook

Introduction

Statistics and Data Science

As of the writing of this book, the fields of statistics and data science are evolving rapidly to meet the changing needs of business, government, and research organizations. It is an oversimplification, but still useful, to think of two distinct communities as you proceed:

1) The traditional academic and medical *research communities* that typically conduct extended research projects adhering to rigorous regulatory or publication standards, and
2) Businesses and large organizations that use statistical methods to extract value from their data, often on the fly. Reliability and value are more important than academic rigor to this *data science community*.

Most users of statistical methods now fall in the second category, as those methods are a basic component of what is now called artificial intelligence (AI). However, most of the specific techniques, as well as the language of statistics, had their origin in the first group. As a result, there is a certain amount of "baggage" that is not truly relevant to the data science community. That baggage can sometimes be obscure or confusing and, in this book, we provide guidance on what is or is not important to data science. Another feature of this book is the use of resampling/simulation methods to develop the underpinnings of statistical inference (the most difficult topic in an introductory course) in a transparent and understandable fashion.

We start off with some examples of statistics in action (including two of statistics gone wrong), then dive right in to look at the proper design of studies and account for the possible role of chance. All the standard topics of introductory statistics are here (probability, descriptive statistics, inference, sampling, correlation, etc.), but sometimes they are introduced not as separate standalone topics but rather in the context of the situation in which they are needed.

Accompanying Web Resources

Python code, datasets, some solutions, and other material accompanying this book can be found at https://introductorystatistics.com/.

Python

Python is a general programming language that can be used in many different areas. It is especially popular in the machine learning and data science communities. A wide range of libraries provide efficient solutions for almost every need, from simple one-off scripts, to web servers, and highly complex scientific applications. As we will see throughout this book, it also has great support for statistics.

You can use Python in many different ways. For most people new to the language, the easiest way to get started is to use Python in Jupyter notebooks (see https://jupyter.org/jupyter.org). Jupyter notebooks are documents that contain both code and rich text elements, such as figures, links, equations, etc. Because of this, they are an ideal environment to learn Python and to present your work. You will find notebooks with the example code of this book on our website (https://introductorystatistics.com/).

A great way to get started with Python is to run code on one of the freely accessible cloud computing platforms. Google Colab (https://colab.research.google.com/) has a free tier that is sufficient for all the examples in this book.

An alternative to cloud computing platforms is to install Python locally on your computer. You can download and install different versions of Python from https://www.python.org. However, it is more convenient to use Anaconda (https://www.anaconda.com). Anaconda is a free package manager for Python and R programming languages focusing on scientific computing. It distributes the most popular Python packages for science, mathematics, engineering, and data analysis. We provide detailed installation instructions on our website at https://introductorystatistics.com/.

Using Python with this Book

With some exceptions, this book presents relevant Python code in the second part of each chapter. The book is not an in-depth step-by-step introduction to computer programming as a discipline, but rather it provides the tools you need to implement the statistical procedures that are discussed in this book. Because many of these procedures are based on iterative resampling, rather than simply calculating formulas, you will get useful practice with the data handling and manipulation that is a Python strength. No specific level of Python ability is required to get started. If you are completely new to Python, you could consider launching yourself with a quick self-study guide (easily found on the web), but, in general, you should be able to follow along.

1

Statistics and Data Science

Statistical methods first came into use before homes had electricity, and had several phases of rapid growth:

- The first big boost came from manufacturers and farmers who were able to decrease costs, produce better products, and improve crop yields via statistical experiments.
- Similar experiments helped drug companies graduate from snake oil purveyors to makers of scientifically proven remedies.
- In the late 20th century, computing power enabled a new class of computationally intensive methods, like the resampling methods that we will study.
- In the early decades of the current millennium, organizations discovered that the rapidly growing repositories of data they were collecting ("big data") could be mined for useful insights.

As with any powerful tool, the more you know about it the better you can apply it and the less likely you will go astray. The lurking dangers are illustrated when you type the phrase "How to lie with..." into a web search engine. The likely autocompletion is "statistics."

Much of the book that follows deals with important issues that can determine whether data yields meaningful information or not:

- How to assess the role that random chance can play in creating apparently interesting results or patterns in data
- How to design experiments and surveys to get useful and reliable information
- How to formulate simple statistical models to describe relationships between one variable and another

We will start our study in the next chapter with a look at how to design experiments, but, before we dive in, let's look at some statistical wins and losses from different arenas.

Statistics for Data Science and Analytics, First Edition. Peter C. Bruce, Peter Gedeck, and Janet Dobbins.

Companion website: www.wiley.com/go/Wiley_Statistics_for_Data

1.1 Big Data: Predicting Pregnancy

In 2010, a statistician from Target described how the company used customer transaction data to make educated guesses about whether customers are pregnant or not. On the strength of these guesses, Target sent out advertising flyers to likely prospects, centered around the needs of pregnant women.

How did Target use data to make those guesses? The key was data used to "train" a statistical model: data in which the outcome of interest—pregnant/not pregnant—was known in advance. Where did Target get such data? The "not pregnant" data was easy—the vast majority of customers are not pregnant, so data on their purchases is easy to come by. The "pregnant" data came from a baby shower registry. Both datasets were quite large, containing lists of items purchased by thousands of customers.

Some clues are obvious—the purchase of a crib and baby clothes is a dead giveaway. But, from Target's perspective, by the time a customer purchases these obvious big ticket items, it was too late—they had already chosen their shopping venue. Target wanted to reach customers earlier, before they decided where to do their shopping for the big day. For that, Target used statistical modeling to make use of non-obvious patterns in the data that distinguish pregnant from non-pregnant customers. One clue that emerged was shifts in the pattern of supplement purchases—e.g. a customer who was not buying supplements 60 days ago but is buying them now.

1.2 Phantom Protection from Vitamin E

In 1993, researchers examining a database on nurses' health found that nurses who took vitamin E supplements had 30% to 40% fewer heart attacks than those who didn't. These data fit with theories that antioxidants such as vitamins E and C could slow damaging processes within the body. Linus Pauling, winner of the Nobel Prize in Chemistry in 1954, was a major proponent of these theories, which were one driver of the nutritional supplements industry.

However, the heart health benefits of vitamin E turned out to be illusory. A study completed in 2007 divided 14,641 male physicians randomly into four groups:

1) Take 268 mg of vitamin E every other day
2) Take 500 mg of vitamin C every day
3) Take both vitamin E and C
4) Take placebo.

Those who took vitamin E fared no better than those who did not take vitamin E. Since the only difference between the two groups was whether or not they took vitamin E, if there were a vitamin E effect, it would have shown up. Several meta-analyses, which are consolidated reviews of the results of multiple published studies, have reached the same conclusion. One found that vitamin E at the above dosage might even increase mortality.

What happened to make the researchers in 1993 think they had found a link between vitamin E and disease inhibition? In reviewing a vast quantity of data, researchers thought they saw an interesting association. In retrospect, with the benefit of a well-designed experiment, it appears that this association was merely a chance coincidence. Unfortunately, coincidences happen all the time in life. In fact, they happen to a greater extent than we think possible.

1.3 Statistician, Heal Thyself

In 1993, Mathsoft Corp., the developer of Mathcad mathematical software, acquired StatSci, the developer of S-PLUS statistical software, the precursor to R. Mathcad was an affordable tool popular with engineers—prices were in the hundreds of dollars and the number of users was in the hundreds of thousands. S-PLUS was a high-end graphical and statistical tool used primarily by statisticians—prices were in the thousands of dollars and the number of users was in the thousands.

In looking to boost revenues, Mathsoft turned to an established marketing principle—cross-selling. In other words, try to convince the people who bought product A to buy product B. With the acquisition of a highly regarded niche product, S-PLUS, and an existing large customer base for Mathcad, Mathsoft decided that the logical thing to do would be to ramp up S-PLUS sales via direct mail to its installed Mathcad user base. It also decided to purchase lists of similar prospective customers for both Mathcad and S-PLUS.

This major mailing program boosted revenues, but it boosted expenses even more. The company lost over $13 million in 1993 and 1994 combined—significant numbers for a company that had only $11 million in 1992 revenue.

What happened?

In retrospect, it was clear that the mailings were not well targeted. The costs of the unopened mail exceeded the revenue from the few recipients who did respond. Mathcad users turned out not to be likely users of S-PLUS. The huge losses could have been avoided through the use of two common statistical techniques:

1) Doing a test mailing to the various lists being considered to (1) determine whether the list is productive and (2) test different headlines, copy, pricing, etc., to see what works best.
2) Using predictive modeling techniques to identify which names on a list are most likely to turn into customers.

1.4 Identifying Terrorists in Airports

Since the September 11, 2001, Al Qaeda attacks in the United States and subsequent attacks elsewhere, security screening programs at airports have become a major undertaking, costing billions of dollars per year in the United States alone. Most of these resources are consumed in an exhaustive screening process. All passengers and their tickets are reviewed, their baggage is screened and individuals pass through detectors of varying sophistication. An individual and his or her bag can only receive a limited amount of attention in a screening process that is applied to everyone. The process is largely the same for each individual. Potential terrorists can see the process and its workings in detail and identify weaknesses.

To improve the effectiveness of the system, security officials have studied ways of focusing more concentrated attention on a small number of travelers. In the years after the attacks, one technique used enhanced screening for a limited number of randomly selected travelers. While it adds some uncertainty to the screening process, which acts as a deterrent to attackers, random selection does nothing to focus attention on high-risk individuals.

Determining who is at high-risk is, of course, the problem. How do you know who the high-risk passengers are?

One method is passenger profiling—specifying some guidelines about what passenger characteristics merit special attention. These characteristics were determined by a reasoned, logical approach. For example, purchasing a ticket for cash, as the 2001 hijackers did, raises a red flag. The Transportation Security Administration trains a cadre of Behavior Detection Officers. The Administration also maintains a specific no-fly list of individuals who trigger special screening.

There are several problems with the profiling and no-fly approaches.

- Profiling can generate backlash and controversy because it comes close to stereotyping. American National Public Radio commentator Juan Williams was fired when he made an offhand comment to the effect that he would be nervous about boarding an aircraft in the company of people in full Muslim garb.
- Profiling, since it does tend to merge with stereotyping and is based on logic and reason, enables terrorist organizations to engineer attackers that do not meet profile criteria.
- No-fly lists are imprecise (a name may match thousands of individuals) and often erroneous. Senator Edward Kennedy was once pulled aside because he supposedly showed up on a no-fly list.

An alternative or supplemental approach is a statistical one—separate out passengers who are "different" for additional screening, where "different" is defined quantitatively across many variables that are not made known to the public. The statistical term is "outlier." Different does not necessarily prove that the person is a terrorist threat, but the theory is that outliers may have a higher threat probability. Turning the work over to a statistical algorithm mitigates some of the controversy around profiling, since security officers would lack the authority to make discretionary decisions.

Defining "different" requires a statistical measure of distance, which we will learn more about later.

1.5 Looking Ahead

We'll be studying many things in this book, but several important themes will be

1) Learning more about random processes and statistical tools that will help quantify the role of chance and distinguish real phenomena from chance coincidence.
2) Learning how to design experiments and studies that can provide more definitive answers to questions such as whether a medical therapy works, which marketing message generates a better response, and which management technique or industrial process produces fewer errors.
3) Learning how to specify and interpret statistical models that describe the relationship between two variables, or between a response variable and several "predictor" variables, in order to:
   - Explain/understand phenomena and answer research questions ("What factors contribute to a drug's success, or the response to a marketing message?").
   - Make predictions ("Will a given subscriber leave this year?" "Is a given insurance claim fraudulent?")

1.6 Big Data and Statisticians

Before the turn of the millennium, by and large, statisticians did not have to be too concerned with programming languages, SQL queries, and the management of data. Database administration and data storage in general was someone else's job, and statisticians would obtain or get handed data to work on and analyze. A statistician might, for example,

- Direct the design of a clinical trial to determine the efficacy of a new therapy
- Help a researcher determine how many subjects to enroll in a study
- Analyze data to prepare for legal testimony
- Conduct sample surveys and analyze the results
- Help a scientist analyze data that comes out of a study
- Help an engineer improve an industrial process

All of these tasks involve examining data, but the number of records is likely to be in the hundreds or thousands at most, and the challenge of obtaining the data and preparing it for analysis was not overwhelming. So the task of obtaining the data could safely be left to others.

1.6.1 Data Scientists

The advent of big data has changed things. The explosion of data means that more interesting things can be done with data, and they are often done in real time or on a rapid turnaround schedule. FICO, the credit-scoring company, uses statistical models to predict credit card fraud, collecting customer data, merchant data, and transaction data 24 hours a day. FICO has more than two billion customer accounts to protect, so it is easy to see that this statistical modeling is a massive undertaking.

The science of computer programming and details of database administration lie beyond the scope of this book, but these fields now lie within the scope of statistical work. The statistician must be conversant with the data, as well as how to get it and work with it.

- Statisticians are increasingly asked to plug their statistical models into big data environments, where the challenge of wrangling and preparing analyzable data is paramount, and requires both programming and database skills.
- Programmers and database administrators are increasingly interested in adding statistical methods to their toolkits, as companies realize that their databases possess value that is strategic, not just administrative, and goes well beyond the original reason for collecting the data.

Around 2010, the term *data scientist* came into use to describe analysts who combined these two sets of skills. Job announcements now carry the term *data scientist* with greater frequency than the term *statistician*, reflecting the importance that organizations attach to managing, manipulating, and obtaining value out of their vast and rapidly growing quantities of data.

We close this chapter with a probability experiment:

Try It Yourself

1) Write down a series of 50 random coin flips without actually flipping the coins. That is, write down a series of 50 made-up H's and T's selected in such a way that they appear random.
2) Now actually flip a coin 50 times.

If you look at the series of made-up tosses and compare them to the real tosses, the longest streaks of either H or T generally occur in the ACTUAL tosses. When a person is asked to make up random tosses, they will rarely "allow" more than four H's or T's in a row. By the time they have written down four H's in a row, they think it is time to switch over to T, or else the series would not appear random. By contrast, instructors who teach this exercise in class often see a streak of 8 T's or H's in a row. Most people think that this is not random, and yet it clearly is.

In 1913, a roulette wheel at the Monte Carlo casino landed on black 26 times in a row. As the streak developed, gamblers, convinced that the wheel would most certainly have to end the streak, increasingly bet heavily on red—they lost millions.

The message here is that random variation reliably produces patterns that *appear* non-random.

Why is this significant? Just as with coin tosses, there is a significant component of random variation (engineers call it noise) in the data that routinely flow through life—whether business life, government affairs, the education world, or personal life. So much data...so much random variation...how do we know what is real and what is random?

We can't know for certain, though we do know that random behavior can appear to be real. One purpose of this book is to teach you about probability, and help you evaluate the potential random component in data, and provide ways of modeling it. This gives us a benchmark against which to measure patterns, and form educated guesses about whether observed events or patterns of interest might be really due to chance. When we understand randomness better, we can curb the tendency to chase after random patterns, and produce more reliable analyses of data.

2

Designing and Carrying Out a Statistical Study

In this chapter, we study random behavior and how it can fool us, and we learn how to design studies to gain useful and reliable information. After completing this chapter, you should be able to

1) Use coin flips to replicate random processes, and interpret the results of coin-flipping experiments
2) Use an informal working definition of probability
3) Define, intuitively, *p*-value
4) Describe the different data formats you will encounter, including relational database and flat file formats
5) Describe the difference between data encountered in traditional statistical research, and "big data"
6) Explain the use of treatment and control groups in experiments
7) Define statistical bias
8) Explain the role of randomization and blinding in assigning subjects in a study
9) Explain the difference between observational studies and experiments
10) Design a statistical study following basic principles

2.1 Statistical Science

"It's not what you don't know that hurts you, it's what you know for sure that ain't so." (Will Rogers, American humorist)

Source: Silvio/Adobe Stock Photos

Statistics for Data Science and Analytics, First Edition. Peter C. Bruce, Peter Gedeck, and Janet Dobbins.

Companion website: www.wiley.com/go/Wiley_Statistics_for_Data

Nearly all large organizations now have huge stores of customer and other data that they mine for insight, in hopes of boosting revenue or reducing costs. In the academic world, over five million research articles are published per year in scholarly and scientific journals. These activities afford ample opportunity to dive into the data, and discover things that aren't true, particularly when the diving is done automatically and at a large scale. Statistical methods play a large role in this extraction of meaning from data. However, the science of statistics also provides tools to study data more carefully, and distinguish what's true from what ain't so.

2.2 Big Data

In most organizations today, raw data are plentiful (often too plentiful), and this is a two-edged sword.

- Huge amounts of data make prediction possible in circumstances where small amounts of data don't help. One type of recommendation system, for example, needs to process millions of transactions to locate transactions with the same item you are looking at—enough so that reliable information about associated items can be deduced.
- On the other hand, huge data flows and incorrect data can obscure meaningful patterns in the data, and generate false ones. Useful data are often difficult and expensive to gather. We need to find ways to get the most information, and the most accurate information, for each dollar spent in assembling and preparing data.

2.3 Data Science

The terms *big data*, *machine learning*, *data science*, and *artificial intelligence* (AI) often go together, and bring different things to mind for different people. The term *artificial intelligence*, particularly with the advent of Chat-GPT and generative AI, suggests almost magical methods that approach human-like cognition capabilities. Privacy-minded individuals may think of large corporations or spy agencies combing through petabytes of personal data in hopes of locating tidbits of information that are interesting or useful. Analysts may focus on statistical and machine learning models that can predict an unknown value of interest (loan default, acceptance of a sales offer, filing a fraudulent insurance claim or tax return, for example).

Statistical science, by contrast, has well over a century of history, and its methods were originally tailored to data that were small and well-structured. However, it is an important contributor to the field of data science which, when it is well practiced, is not just aimless trolling for patterns, but starts out with questions of interest:

- What additional product should we recommend to a customer?
- Which price will generate more revenue?
- Does the MRI show a malignancy?
- Is a customer likely to terminate a subscription?

All these questions require some understanding of random behavior and all benefit from an understanding of the principles of well-designed statistical studies, so this is where we will start.

2.4 Example: Hospital Errors

Healthcare accounts for about 18% of the United States GDP (as of 2024), is a regular subject of political controversy and proposals for reform, and produces enormous amounts of data and analysis. One area of study is the problem of medical errors—violations of the Hippocratic oath's "do no harm" provision. Millions of hospitalized patients each year around the world are affected by treatment errors (mostly medication errors). A 2017 report from the National Institutes of Health (NIH) in the U.S. estimated that 250,000 deaths per year resulted from medical errors. There are various approaches to dealing with the problem.

Source: Vineey/Adobe Stock Photos

Clinical Decision Support systems (CDS) are used to guide practitioners in diagnosis and treatment, and can provide rule-based alerts when standard treatment protocols are violated. However, all those rules must be programmed and kept up-to-date in an extremely complex medical environment. Many false alarms result, which can cause practitioners to ignore the alerts. Recent advances in machine learning have enabled systems that learn on their own to provide alerts, without experts having to program rules. These systems allow for the *correction* of errors once they occur, but what about identifying the *causes* of errors and reducing their frequency?

One obvious and uncontroversial innovation has been to promote the use of checklists to reduce errors of omission. Other ideas may not be so obvious. No-fault error reporting has been proposed, in which staff are encouraged to report all errors, both their own and those committed by others, without fear of punishment. This could have the benefit of generating better information about errors and their sources, but could also hinder accountability efforts. How could you find out whether such a program really works? The answer: a well-designed statistical study.

2.5 Experiment

To tie together our study of statistics we will look at an experiment designed to test whether no-fault reporting of *all* hospital errors reduces *major* errors in hospitals (errors resulting in further hospitalization, serious complications, or even death). An experiment like this was conducted by a hospital in Quebec, Canada, but it was too small to provide definitive conclusions. For illustrative purposes, we will look at hypothetical data that a larger study might have produced.

Experiments are used in industry, medicine, social science, and data science. The ubiquitous A/B test (more on that later) is an experiment. The key feature of an experiment is that the investigator *manipulates* some variable that is believed to affect an outcome of interest, in order to demonstrate the importance and effect (or lack thereof) of the variable. This stands in contrast to a survey or other analysis of existing data, where the analyst simply collects and analyzes data. For example, in a web experiment, the marketing investigator might try out a new product price to see how it affects sales.

Experiments can be uncontrolled or controlled. In an uncontrolled experiment, the investigator collects data on the group or time period for which the variable of interest has been changed. In a web experiment, for example, the price of a product might be increased by 25%, and then sales compared to prior sales.

Experiment vs. Observational Study

In the fifth inning of the third game of the 1932 baseball World Series between the NY Yankees and the Chicago Cubs, the great slugger Babe Ruth came to bat and pointed towards center field, as if to indicate that he planned to hit the next pitch there. On the next pitch, he indeed hit the ball for a home run into the center field bleachers.[a]

A Babe Ruth home run was an impressive feat, but not that uncommon. He hit one every 11.8 at-bats. What made this one so special is that he predicted it. In statistical terms, he specified *in advance* a theory about a future event—the next swing of the bat—and an outcome of interest—a home run to center field.

In statistics, we make an important distinction between studying pre-existing data (an observational study) and collecting data to answer a pre-specified question (an experiment or prospective study). The most impressive and durable results in science come when the researcher specifies a question in advance, then collects data in a well-designed experiment to answer the question. Offering commentary on the past can be helpful, but is no match for predicting the future.

[a]There is some controversy about whether he actually pointed to center field or to left field and whether he was foreshadowing a prospective home run or taunting Cubs players. You can Google the incident ("Babe Ruth called shot") and study videos on YouTube, then judge for yourself.

2.6 Designing an Experiment

The problem with an uncontrolled experiment is the uncertainty involved in the comparison. Suppose sales drop 10% in the web experiment with the new price. Can you be sure that nothing else has changed since the experiment started? Probably not—companies are making modifications and trying new things all the time. Hence, the need for a control group.

In a controlled experiment, two groups are used and they are handled in the same way, except that one is given the treatment (e.g. the increased price), and the other is not given the treatment. In this way, we can eliminate the confounding effect of other factors not being studied.

2.6.1 A/B Tests; A Controlled Experiment for the Hospital Plans

In our errors experiment, we could compare two groups of hospitals. One group uses the no-fault plan and one does not. The group that gets the change in treatment you wish to study (here, the no-fault plan) is called the *treatment group*. The group that gets no treatment or the standard treatment is called the *control group*.

An experiment like this, testing a control group vs. a treatment group, is also called an A/B test, particularly in the field of marketing, where one web treatment might be tested against another. Sometimes, particularly in marketing, there might not be an established control scenario and we are simply comparing one proposed new treatment against another proposed new treatment (e.g. two different web pages).

How do you decide which hospitals go into which group?

You would like the two groups to be similar to one another, except for the treatment/control difference. That way, if the treatment group does turn out to have fewer errors, you can be confident that it was due to the treatment. One way to do this would be to study all the hospitals in detail, examine all their relevant characteristics, and assign them to treatment/control in such a way that the two groups end up being similar across all these attributes. There are two problems with this approach.

1) It is usually not possible to think of all the relevant characteristics that might affect the outcome. Research is replete with the discovery of factors that were unknown prior to the study or thought to be unimportant.
2) The researcher, who has a stake in the outcome of the experiment, may consciously or unconsciously assign hospitals in a way that enhances the chances of the success of their pet theory.

Oddly enough, the best strategy is to assign hospitals randomly: for example, by tossing a coin.

2.6.2 Randomizing

True random assignment eliminates both conscious and unconscious bias in the assignment to groups. It does not guarantee that the groups will be equal in all respects. However, it does guarantee that any departure from equality will be due simply to the chance allocation, and that the larger the samples, the fewer differences the groups will have. With extremely large samples, differences due to chance virtually disappear, and you are left with differences that are real—provided the assignment to groups is really random.

Random assignment lets us make the claim that any difference in group outcomes that is *more* than might reasonably happen by chance is, in fact, due to the different treatment of the groups. The study of probability in this book lets us quantify the role that chance can play and take it into account.

We can imagine an experiment in which *both* groups got the same treatment. We would expect to see some differences from one hospital to another. An everyday example of this might be tossing a coin. If you toss a coin 10 times you will get a certain number of heads. Do it again and you will probably get a different number of heads.

Though the results vary, there are laws of chance that allow you to calculate things like how many heads you would expect on average or how much the results would vary from one set of 10 (or 100 or 1000) tosses to the next. If we assign subjects at random, we can use these same laws of chance—or a lot of coin tosses—to analyze our results.

If we have Doctor Jones assign subjects using her own best judgment, we will have no knowledge of the (often subconscious) factors that influence assignment. These factors may bias assignment so that we can no longer say that the *only* thing (besides chance assignment) distinguishing the treatment and control groups is the treatment. Random assignment is not always possible—for example, randomly assigning elementary school students to two different teaching methods, where everything else in the education setting is the same.

Randomization can be difficult or impossible in some situations, but it is relatively easy in the A/B testing that is popular in digital marketing. Web visitors can be easily randomized to one web page or another; email recipients can easily be assigned randomly to one version or another of an email.

2.6.3 Planning

You need some hospitals and you estimate you can find about 100 within a reasonable distance. You will probably need to present a plan for your study to the hospitals to get their approval. That seems like a nuisance, but they cannot let just anyone do any study they please on the patients.[1] In addition to writing a plan to get approval, you know that one of the biggest problems in interpreting studies is that many are poorly designed. You want to avoid that, so you think carefully about your plan and ask others for advice. It would be good to talk to a statistician with experience in medical work. Your plan is to ask the 100 or so available hospitals if they are willing to join your study. They have a right to say no. You hope quite a few will say yes. In particular, you hope to recruit 50 willing hospitals and randomly assign them to treatment and control.

> **Try It Yourself**
> How exactly would you assign hospitals randomly? Think about options for the scenario where it doesn't matter if the groups are exactly equal-sized, and for the scenario where you want two groups of equal size.

1 This effort is modest, in comparison to that required to gain approval for new drug therapies. Clinical trials to establish drug efficacy and safety can take years and cost billions of dollars.

Figure 2.1 Dart throws off-target in *consistent* fashion (biased). Source: Peter Bruce (Book Author).

2.6.4 Bias

Randomization is used to try to make the two groups similar at the beginning. It is important to keep them as similar as possible during the experiment. We want to be sure the treatment is the only difference between them. Any difference in outcome due to non-random extraneous factors is a form of *bias*. Statistical bias is a technical concept but can include the lay definition that refers to people's opinions or states of mind.

Definitions: Bias Statistical bias is the tendency for an estimate, model, or procedure to yield results that are consistently off-target for a specific purpose (as in Fig. 2.1).

For example, the mean (average) income for a region might not be a good estimate for the income of a typical resident, if part of the region is home to a small number of very high-income residents. Their incomes would likely raise the average above that of most typical residents (i.e. ones selected at random). Another example is gun sights on a long-range rifle. Gravity will exert a downward pull on a bullet, the more distant the target the greater the pull. The coordinates of the target in the sights will be biased upward compared to where the bullet lands.

Bias can often creep into a study when humans are involved, either as subjects or experimenters. For one thing, subject behavior can be changed by the fact that they are participating in a study. Experience has also shown that people respond positively to attention, and just being part of a study may cause subjects to change. A positive response to the attention of being in a study is called the *Hawthorne effect*. Awareness of an issue can significantly affect perceptions, which is why potential jurors in a trial are asked if they have seen news coverage of a case at issue.

Out-of-Control Toyotas?

In the fall of 2009, the National Highway Transportation Safety Agency (NHTSA) was receiving several dozen complaints per month about Toyota cars speeding out of control. The rate of complaint was not that different from the rates of complaint for other car companies. Then, in November of 2009, Toyota recalled 3.8 million vehicles to check for sticking gas pedals. By February, the complaint rate had risen from several dozen per month to over 1500 per month of alleged cases of unintended acceleration. Attention turned to the electronic throttle.

Clearly what changed was not the actual condition of cars—the stock of Toyotas on the road in February of 2010 was not that different from November of 2009. What changed was car owners' awareness and perception as a result of the headlines surrounding the recall. Acceleration problems, whether real or illusory, that escaped notice prior to November 2009 became causes for worry and a trip to the dealer. Later, the NHTSA examined a number of engine data recorders from accidents where the driver claimed to experience acceleration despite applying the brakes. In all cases, the data recorder showed that the brakes were not applied.

In February 2011, the US Department of Transportation announced that a 10-month investigation of the electronic throttle showed no problems. Public awareness of the problem boosted the rate of complaint far out of proportion to its true scope.

Lesson: Your perception of whether you personally experience a problem or benefit is substantially affected by your prior awareness of others' problems and benefits.

Sources: *Wall Street Journal*, July 14, 2010; The Analysis Group (http://www.analysisgroup.com—accessed July 14, 2010); *USA Today* online, April 2, 2011.

In some situations, we can avoid telling people they are participating in a study. For example, a marketing study might try different ads or products in various regions without publicizing that they are doing so for research purposes. In other situations, we may not be able to avoid letting subjects know they are being studied, but we may be able to conceal whether they are in the treatment or control group.

2.6.4.1 Placebo

A placebo is a dummy treatment imposed on the control group to render their experience similar to that of the control group. In this way, we can distinguish the real effect of the treatment from any response that is simply the result of perceiving that you are being treated. Experience has shown that subjects will often experience and report good results even for dummy treatments. This positive response to the perception that you are being treated is called the *placebo effect*. In medicine, the placebo effect on the brain can be powerful for mitigating symptoms, and explains the popularity of many remedies that have no rigorous scientific basis for their effectiveness. In one study, a tablet labeled "placebo" was found to be 50% as effective as actual migraine medicine in relieving migraine symptoms. Thus, the net beneficial effect of many therapies is made up both of a real scientific component, and a placebo component.

2.6.4.2 Blinding

The process of concealing treatment from control is termed *blinding*. We say a study is *single-blind* when the subjects—the hospitals in our medical errors example—do not know whether they are getting the treatment. It is *double-blind* if the staff in contact with the subjects also do not know which group is getting the real treatment. It is *triple-blind* if the people who evaluate the results do not know, either.

Blinding, particularly full blinding, is not always feasible. The very nature of the treatment (for example, no-fault error reporting for the hospital study) may require participant awareness. The control side may need to be more than simply "do nothing." Drug trials typically include a sugar pill for the control arm. For the hospital study, our control might be a standard set of best practices (excluding no-fault reporting) shared among the control hospitals.

In marketing experiments, participant blinding usually happens automatically—web viewers or message recipients are typically unaware there might be another group seeing a different message. Bias from the analyst side, though, is common. It often occurs when analysts have a favored outcome (perhaps subconsciously) and can choose when to end an experiment, or elect to look at a subset of results that seem more reasonable to them. Blinding of the analyst can mitigate this bias.

In addition to the various forms of blinding, we try to keep all other aspects of each subject's environment the same. In the hospital experiment, we need to agree on common treatment and control error-handling methods that will be applied to all hospitals in each group. By keeping the two groups the same in every way except the treatment, we can be confident that any differences in the results were due to it.

2.6.4.3 Before-after Pairing

We could run our study for a year and measure the total number of medical errors each hospital had by the end of that period. A better strategy is to measure how many errors they had the year before the study as well. Then we have *paired data*—two measurements on each unit. This allows us to compare the treatment to no treatment *at the same hospitals*.

Note that we still retain the control group. Having both a control group and a treatment group allows us to separate out the improvement due to no-fault-reporting from the improvement due to the more general best practices treatment. Having a control group also controls for trends that affect all hospitals. For example, the number of errors could be increasing due to an increased patient load at hospitals generally.

2.7 The Data

Let's look at the results—major errors in the year preceding the treatment, and during the period when some hospitals implemented the treatment intervention (the no-fault reporting plan). A "0" in the treatment column in Table 2.1 indicates the hospital was not selected for the no-fault treatment program, a "1" indicates it was.

Let's now focus directly on the subject of interest—reduction in errors. Table 2.2 shows the full table, listing only the amount by which errors were reduced.

2.7.1 Dataframe Format

Tables 2.1 and 2.2 are examples of a standard database or tabular format, which all database programs and most standard-purpose statistical software programs use.

Table 2.1 Hospital errors (partial): before and after treatment.

Row	Hospital#	Treat?	Errors Before	Errors After
1	239	0	27	24
2	1126	0	17	16
3	1161	0	31	29
4	1293	1	38	32
5	1462	1	25	23
	45 more			

Table 2.2 Reduction in major errors in hospitals.

Row	Hospital#	Treat?	Reduction in Errors	Row	Hospital#	Treat?	Reduction in Errors
1	239	0	3	26	2795	1	3
2	1126	0	1	27	2889	0	5
3	1161	0	2	28	2892	1	9
4	1293	1	2	29	2991	1	2
5	1462	1	2	30	3166	1	2
6	1486	0	2	31	3190	0	1
7	1698	1	5	32	3254	0	4
8	1710	0	1	33	3312	1	2
9	1807	0	1	34	3373	1	2
10	1936	1	2	35	3403	1	3
11	1965	1	2	36	3403	0	1
12	2021	1	2	37	3429	1	2
13	2026	0	1	38	3441	1	6
14	202	0	3	39	3520	0	1
15	208	1	4	40	3568	1	2
16	2269	1	2	41	3580	0	2
17	2381	1	2	42	3599	0	2
18	2388	0	1	43	3660	1	2
19	2400	1	2	44	3985	0	2
20	2475	0	4	45	4014	1	2
21	2548	0	1	46	4060	0	1
22	2551	0	2	47	4076	1	2
23	2661	0	1	48	4093	0	1
24	2677	1	4	49	4230	0	2
25	2739	1	2	50	5633	0	2

The standard object in data programming languages is a *dataframe*. Rows represent records or cases—hospitals in this example. Columns represent variables, which are data that change from case to case (hospital to hospital). The format has two key features:

1) Each row contains all the information for one and only one case.
2) All data for a given variable is in a single column.

To load data from a .csv file[2] into a dataframe named "data," you would use `pandas` with the syntax `data = pd.read_csv("hospitalerrors.csv")`. We will cover this in more detail in Section 2.19.3.

Let's look at the hospital data.

Column 1 is simply the row number.

Column 2, hospital, contains *case labels*. These are arbitrary labels for the experimental units—a unique number for each unit. Case labels keep track of the data. For example, if we find a mistake in the data, we would need to know which hospital that came from so we could investigate the cause and correct the mistake. Numerical codes are preferred to more informative labels when we wish to conceal the group to which subjects were assigned.

Column 3 labels observations from the treatment group with a one and those from the control group with a zero.

Column 4 is the number of major medical errors in the year before the study minus the number from the following year. A positive number represents a reduction in medical errors. Note that all the numbers are positive—things got better whether subjects got the treatment or not! This could be due to the Hawthorne effect, or to any extra care the subjects got from being in the experiment, or to other factors that may have changed globally between the two years.

2.8 Variables and Their Flavors

2.8.1 Numeric Variables

The third (*Treat?*) and fourth (*Reduction in errors*) columns in Table 2.2 contain *variables*. These are things we observe, compute, or measure for each subject. Each row represents an experimental unit or subject, while each column represents a variable. Note that we have gone from listing the number of errors at the beginning of the study and the number at the end to simply listing the reduction in errors, which is what we are really interested in. They are an example of a *numeric* variable, also called a *quantitative variable*. These are numbers with which you can do meaningful arithmetic, and typically measure magnitudes—how much of something. (The hospital error data is a special type of numerical variable—it is "count" data.)

2 "Comma separated values," or .csv, is a simple file format that can be read by a wide variety of programming languages and software.

2.8.2 Categorical Variables

The other main type of variable is called *categorical*, or sometimes *factor*. Examples that might be in the database (though not printed out above) are the city, county, or province of each hospital, or whether it was a government, business, or charity hospital. Categorical data is often recorded in text labels, for example: Male or Female, Christian, Muslim, Hindu, Buddhist, Jew, or Other. But it is also common to code categories numerically. In the hospital data, treatment is a categorical variable and was coded as one. Control is coded as zero.

A categorical variable must take one of a set of defined non-numerical values—yes/no, low/medium/high, mammal/bird/reptile, etc. The categories might be represented as numbers to accommodate software, so you need to be cautious about doing routine arithmetic on those numbers, and be aware of what the arithmetic really means. The choice of numbers is usually arbitrary.

2.8.3 Binary Variables

A special type of categorical variable is a *binary*, or two-value, variable. Much data has a binary variable as an outcome, even some data that starts out as multi-category outcome. Examples are survive or die, purchase or no-purchase, click or no-click, fraud or no-fraud. The prevalence of binary or yes-no data reflects the fact that decision-making is often easier if situations, even complex ones, can be boiled down to a choice between two alternatives. Binary data is typically represented by 1's and 0's, with 1 representing the more unusual and interesting category.

Coding categories as numbers dose not make them numeric! Don't do arithmetic on numerical codes for categorical data when it makes no sense. If we code Christian, Muslim, Hindu, Buddhist, Jew, or Other as 1, 2, 3, 4, 5, 6, respectively, then finding the total or average of these codes is not meaningful. When the categories are ordered, such as the degree of pain, some limited calculations may be possible.

With binary data in which one class is much more scarce than the other (e.g. fraud/no-fraud), the scarce class is often designated "1." Surveys and observational studies often produce categorical data. People might be asked which candidate they plan to vote for; what toothpaste they use; or whether they live in a house, apartment, or mobile home. Observational studies often use similar variables that can be evaluated at a glance or found in existing records. The basic statistical summary method for displaying such categorical data is a *frequency table* (see Table 3.5 for an example).

With 0/1 data, for example, "fraudulent" or "OK," the more rare and interesting category (e.g. a fraudulent credit card charge) is designated "1." Source: energepic.com/Pexels.

2.8.4 Text Data

Another type of data encountered is *text* data. Text data can appear in a wide variety of formats: ordinary prose, posts on social media, a doctor's notes, the contents of labels, etc. Not usually thought of as a variable, text data typically must be preprocessed before it can be subjected to statistical analysis, although some machine learning methods incorporate the needed preprocessing and can deal with text data directly.

2.8.5 Random Variables

A variable that takes on different values (e.g. heads/tails) as the result of a random process is a *random variable.*

The distinction between what is random and what is not is often fuzzy. A coin flip, you would think, is purely unpredictable, yet magicians are able to flip coins so that they always appear to land heads. The quantity of acetaminophen in an extra strength Tylenol you would consider to be invariant at 500 mg, yet there is always some tiny uncontrolled variability in the exact amount of acetaminophen per tablet. If we drill down to a sufficiently fine level of resolution, we usually encounter an element of randomness in most measurements.

2.8.6 Simplified Columnar Format

An alternative is to present the error reduction for the control group in one column and the treatment group in the other (see Table 2.3). This provides the clearest presentation of how the two groups differ in the extent to which errors were reduced. The treatment group had, on average, 2.80 fewer errors in the second year.

Table 2.3 Error reduction: compact table.

	Treatment	Control
	2	3
	2	1
	5	2
	2	2
	2	1
	2	1
	4	1
	2	3
	2	1
	2	4
	4	1
	2	2
	3	1
	9	5
	2	1
	2	4
	2	1
	2	1
	3	2
	2	2
	6	2
	2	1
	2	1
	2	2
	2	2
Mean:	2.80	1.88

The control group had 1.88 fewer errors in the second year. Both groups reduced errors, but the treatment does appear to have reduced errors to a greater extent than the control.

2.9 Python: Data Structures and Operations

Data structures and their manipulation are key features of every programming language. This section introduces standard data types and more complex data structures as well as some basic operations to manipulate them.

2.9.1 Primary Data Types

Python has the following primary data types:

- **int**: Represents integer values, such as 1, −5, or 1000.
- **float**: Represents floating-point or decimal values, such as 3.14, −0.5, or 2.0.
- **string**: Represents sequences of characters enclosed in single (`'`) or double quotes (`"`). For example, `"Hello, World!"` or `'Python'`.
- **bool**: Represents the boolean values, `True` and `False`.

The Python standard library also has support to handle data describing events at a specific time and/or date:

- **time**: Represents time values, e.g. 8 AM or 13:20.
- **date**: Represents dates, e.g. May 4th, 2023.
- **datetime**: Represents date and time together, e.g. 8 AM on May 4th, 2023.

We are not going to cover these three data types in this book. A special data type is `None` which is representing the absence of a value.

You can use the *type()* function to check the data type of a value.

```
print(type(3.14))      # <class 'float'>
print(type("Python")) # <class 'str'>
print(type(True))      # <class 'bool'>
```

2.9.2 Comments

Every programming language has a way to add comments to code. Comments are used to add information about the code, but will not be executed. Comments are started with a hash (#) and continue until the end of the line.

```
# This is a comment
print("Hello, World!")  # This is also a comment
print("Inside a string # this is not a comment")
```

In the third line, the # character is inside a string, so it is not interpreted as a comment. If we execute this code, we will get the following output.

```
Hello, World!
Inside a string # this is not a comment
```

2.9.3 Variables

Variables are used to store values in Python ("variables" in the computer science world are related to, but different from "variables" in the statistical sense). In Python, variables are words that start with a letter or underscore (_) and can contain more letters, numbers, and underscores. Variable names are case-sensitive, so `message` and `Message` are two different variables. There are some reserved words that cannot be used as variable names. You can assign a value to a variable using the assignment operator (=).

```
pi_value = 3.14
message = "Python"
is_raining = True
_formula_1 = "H2O"
```

You can view a variable as a synonym for the actual value and use it in place of the value.

```
print(pi_value)      # output: 3.14
```

Variables can be reassigned new values as many times as needed.

```
pi_value = 3.14          # value is 3.14
pi_value = 3.14159       # value is now 3.14159
two_pi = 2 * pi_value    # value of two_pi is now 6.28318
```

2.9.4 Operations on Data

Variables by itself are useful to keep information, but the real value of variables is that you can perform operations on them. We've already seen an example above, where we used the `pi_value` to calculate its double value. This and other numerical operations in Python allow you to perform calculations on numerical data types, such as integers and floats. You can use arithmetic operators like addition (+), subtraction (-), multiplication (*), division (/), potentiation (**) and modulus (%).

```
print(2 + 3)               # output: 5
print(2.0 - 3)             # output: -1.0
print(10 * 3 / 6 + 4)      # output: 9.0
print(2 ** 3)              # output: 8 (2 to the power of 3)
print(10 * 3 / (6 + 4))    # output: 3.0
```

```
# Modulus operator returns the remainder of the division
print(8 % 3)            # output: 2
print(8.0 % 3)          # output: 2.0
```

Strings can be combined using the addition operator (+) and repeated using the multiplication operator (*).

```
print("Hello, " + "World!")  # output: Hello, World!
print("Python " * 3)         # output: Python Python Python
```

If you want to access individual characters or substrings, you can index and slice using square brackets ([]).

```
message = "Python"
print(message[0])      # output: P
print(message[0:2])    # output: Py
print(message[2:])     # output: thon
```

Strings also have a number of useful methods for common operations like replacing substrings, converting to uppercase or lowercase, and splitting into substrings.

```
message = "Hello, World!"
print(message.replace("World", "Python"))  # output: Hello, Python!
print(message.upper())                     # output: HELLO, WORLD!
print(message.lower())                     # output: hello, world!
print(message.split(","))                  # output: ['Hello', ' World!']
```

2.9.4.1 Converting Data Types

Sometimes, it becomes necessary to convert data from one type to another. Python provides functions like *int()*, *float()*, and *str()* to convert values between integer, float, and string data types, respectively. This subsection explores the conversion of data types and their significance in different scenarios.

Consider for example that you read a file, so any number will be presented as a string of digits, for example `"4213"`. Using the *int* function, we can convert it to a number. If the text cannot be converted correctly, an exception is shown. The same works for *float* and text representing real numbers like `"3.1415"`.

```
text = "4213"
print(int(text))  # output: 4213
text = "3.1415"
print(float(text))  # output: 3.1415
```

The reverse is done by the *str* function. In fact, it converts any value to a string, although not all might be very informative. This is useful if you want to show the result of a calculation in a report or on the screen. The *print* function does this actually automatically.

```
import math
print(math.pi)  # output: 3.141592653589793
```

The result might be good enough for a preliminary analysis, most of the time, we want to control the output more. For example, we might want to only print the first two digits after the decimal point and combine the output with a string. There are a variety of ways to create formatted output in Python. In our opinion, the most flexible and readable way is to use *f-strings*.

```
print(f"PI = {math.pi:.4f}")  # output: PI = 3.1416
```

The `f` in front of the string indicates that it is a so-called *f-string*. The curly brackets are used to indicate that the value of a variable or expression should be inserted at this position. Here, the expression is `math.pi`. The part after the colon defines how the values are displayed. The `.4f` specifies that we print a *float* value and that we want to show four digits after the period. You can find more information about format specifiers in the Python documentation.[3]

2.9.5 Advanced Data Structures

Apart from the basic data types, Python offers advanced data structures that allow you to organize and store more complex data. This subsection focuses on four essential data structures:

list: A list is an ordered collection of elements of any type, e.g. `[1, 2, "a", [4, 5]]`.

tuple: A tuple is similar to a list, but cannot be modified after creation, this means tuples are immutable, e.g. `(1, 2, "a")`.

set: A set is an unordered collection of unique elements, e.g. `{1, 2, 3}`.

dict: A dictionary is a collection of key-value pairs, e.g. `{"a": 1, "b": 2}`.

Lists are created using the *list* function or using square brackets (`[]`). In the example, we create a list of integers and an empty list. We can get the length of the list using the *len* function.

```
list1 = [0, 1, 2, 3, 4, 5, 6]
list2 = []
print(len(list1))  # output: 7
print(len(list2))  # output: 0
```

You can access individual elements of the list by their position. The position is usually referred to as the *index*. The index of the first element is 0, the index of the second element is 1, and so on. You can also use negative indices to access elements counting from the end of the list. The index of the last element is −1, the index of the second last element is −2, and so on.

3 https://docs.python.org/3/library/string.html#format-specification-mini-language

```
print(list1[0]) # output: 0
print(list1[1]) # output: 1
print(list1[-1]) # output: 6
print(list1[-2]) # output: 5
```

Lists also allow slices to access a continuous sequence of elements.

```
print(list1[1:3]) # output: [1, 2]
print(list1[2:])  # output: [2, 3, 4, 5, 6]
print(list1[:3])  # output: [0, 1, 2]
```

You can modify the elements of a list by assigning new values to the elements, appending or inserting new values, or removing existing values.

```
list1[0] = 10
print(list1)          # output: [10, 1, 2, 3, 4, 5, 6]
list1.append(7)
print(list1)          # output: [10, 1, 2, 3, 4, 5, 6, 7]
list1.insert(3, 11) # insert 11 at index 3
print(list1)          # output: [10, 1, 2, 11, 3, 4, 5, 6, 7]
list1.remove(10)      # remove first occurrence of 10 from the list
print(list1)          # output: [1, 2, 11, 3, 4, 5, 6, 7]
list1.pop(2)          # remove element at index 2
print(list1)          # output: [1, 2, 3, 4, 5, 6, 7]
```

Tuples are created using the *tuple* function or using parentheses (`()`). They are similar to lists, but cannot be modified after creation.

```
tuple1 = (1, 2, 3, 4, 5, 6)
tuple2 = ()
print(len(tuple1))  # output: 6
print(len(tuple2))  # output: 0
print(tuple1[0])    # output: 1
print(tuple1[1:3])  # output: (2, 3)
```

You cannot replace, append, or remove values. The *namedtuple* function from the `collections` module is a more flexible alternative to the standard tuple. You can use it to create a tuple with a fixed set of fields, where the individual fields can be accessed by names and not only by index. This is very useful in larger projects as it makes the code more readable and less error-prone.

```
from collections import namedtuple
Point = namedtuple("Point", ["x", "y"]) ①
p = Point(1, 2) ②
print(p.x)  # output: 1
print(p.y)  # output: 2
```

① This line creates a *namedtuple* called `Point` with the fields `x` and `y`.

② This line creates a `Point` object with the values 1 and 2 for the fields `x` and `y`. A slightly more readable alternative is to explicitly state the field names, `Point(x=1, y=2)`.

An alternative way of defining named tuples is the *NamedTuple* class from the `typing` package (see Section 2.9.5.2). It is the method of choice to define named tuples with type information.

```
from typing import NamedTuple
class Point(NamedTuple):
    x: int
    y: int
```

We are not using it in this book, but you will find it useful in your own projects.

Sets are similar to lists, but represent an unordered collection of unique elements. Sets are created using the *set* function or using curly braces ({}).

```
set1 = {1, 2, 3, 4, 5, 6}
set2 = set()
print(len(set1))  # output: 6
print(len(set2))  # output: 0
```

You can add new elements to a set using the *add* method and remove elements using the *remove* method.

```
set1.add(7)
print(set1)  # output: {1, 2, 3, 4, 5, 6, 7}
set1.remove(7)
print(set1)  # output: {1, 2, 3, 4, 5, 6}
```

We can also combine two sets in different ways. The *union* method returns a new set with all elements from both sets, the *intersection* method returns a new set with elements that are common to both sets, and the *difference* method returns a new set with elements in the first set but not in the second set.

```
set1 = {1, 2, 3, 4, 5, 6}
set2 = {4, 5, 6, 7, 8, 9}
print(set1.union(set2))          # output: {1, 2, 3, 4, 5, 6, 7, 8, 9}
print(set1.intersection(set2))   # output: {4, 5, 6}
print(set1.difference(set2))     # output: {1, 2, 3}
print(set2.difference(set1))     # output: {8, 9, 7}
```

There are short hand versions of these methods using the operators | for union, & for intersection, and - for difference.

```
print(set1 | set2)  # output: {1, 2, 3, 4, 5, 6, 7, 8, 9}
print(set1 & set2)  # output: {4, 5, 6}
print(set1 - set2)  # output: {1, 2, 3}
print(set2 - set1)  # output: {8, 9, 7}
```

Sets can also be useful if you want to remove duplicate elements from a list.

```
list1 = [1, 2, 3, 4, 5, 6, 1, 2, 3]
print(list1)  # output: [1, 2, 3, 4, 5, 6, 1, 2, 3]
list1 = list(set(list1))
print(list1)  # output: [1, 2, 3, 4, 5, 6]
```

In lists, we identified a specific element by the position in the list. But what would we use if we want to lookup US state names (e.g. Virginia) using their abbreviation (e.g. VA). A dictionary is the perfect data structure for this task. Here is an example:

```
states = {"VA": "Virginia", "MD": "Maryland", "DC": "District of Columbia"}
print(states["VA"])  # output: Virginia
print(states["MD"])  # output: Maryland
```

The dictionary is defined as a comma-separated list of key-value pairs. The key and the value are separated by a colon (`:`). The key is used to lookup the value. You can add new key-value pairs to a dictionary using the assignment operator (=) and remove key-value pairs using the *del* function.

```
states["NY"] = "New York"
print(states)  # output: {"VA": "Virginia", "MD": "Maryland",
               #          "DC": "District of Columbia", "NY": "New York"}
del states["NY"]
print(states)  # output: {"VA": "Virginia", "MD": "Maryland",
               #          "DC": "District of Columbia"}
```

You can find many more specialized data structures in the Python standard library. One that is useful to know is the `collections` module. It contains a number of specialized data structures like *namedtuple*, *Counter*, and *defaultdict*. We already learned about *namedtuple* and will cover *Counter* in Section 6.7. The *defaultdict* is a dictionary subclass that calls a factory function to supply missing values. This sounds more complex than it is. Here is an example:

```
from collections import defaultdict
word_counts = defaultdict(int)  ①
for word in ["apple", "banana", "apple", "banana", "apple"]:
    word_counts[word] += 1  ②
print(word_counts)  # output: defaultdict(<class 'int'>, {'apple': 3,
   'banana': 2})
```

① This line creates a *defaultdict* called `word_counts` with the default value of `int`. This means that if a key is not found, the value will be initialized with `0`. Other types are also possible, for example `list` or `set`. In this case, the default value will be an empty list or set.

② This line increments the value of the key `word` by 1. If the key is not found, the default value of `0` is used. This means `word_counts[word]` is initialized with `0` and then incremented by 1. The operator += is a shorthand for `word_counts [word] = word_counts[word] + 1`.

As the example shows, the *defaultdict* is useful if you want to count the occurrences of elements in a list to create a frequency table (see Section 3.6.1).

Later on, you will learn about `pandas` and `numpy`. These two packages are at the core of many statistical packages in Python which require efficient implementation of data tables and numerical arrays. For details see Section 2.19.

2.9.5.1 Classes and Objects

The data structures we covered so far, allow us to represent one or more pieces of information. Classes and objects take this idea one step further. They combine data and functions that operate on these data. The difference between class and object is that *class* refers to the definition or implementation, while an *object* is an instance of a class.

Here is an example. We want to capture information about a person. A `Person` class might have a first name, family name, and a birth date. It might also have a method (a function that is associated with a class) to calculate the full name.

```
class Person:  ①
    def __init__(self, first_name, family_name, birth_date):  ②
        self.first_name = first_name  ③
        self.family_name = family_name
        self.birth_date = birth_date

    def full_name(self):  ④
        return f"{self.first_name} {self.family_name}"  ⑤
```

① This line indicates that we want to create a class called `Person`.
② This line means that we store the value of the `first_name` argument in the field `self.first_name` of the object.
③ This function is called an initializer. In order to create a `Person` object, we need to provide the first name, family name, and birth date. The `self` parameter is a reference to the object itself. We use it to store the values in the object.
④ `full_name` is a method of the `Person` class. The `self` parameter is a reference to the object itself. We use it to access the values stored in the object.
⑤ `self.first_name` and `self.family_name` are the values stored in the object identified by the `self` parameter. We use them to create the full name with an f-string.

This might sound complicated, but it is actually quite simple. Let's create a `Person` object and use it.

```
person = Person("John", "Doe", "1970-01-01")  ①
print(person.first_name)  # output: John  ②
print(person.full_name())  # output: John Doe  ③
```

① This line creates a `Person` object with the first name `John`, family name `Doe` and birth date `1970-01-01`.
② We can always access the fields of an object directly using the dot operator (`.`).
③ This line calls the `full_name()` method of the person object and prints the result.

We can have as many instances of a class as we want. Each instance is independent of the others. Let's add a second person to our example and store both in a list.

```
person1 = Person("John", "Doe", "1970-01-01")
person2 = Person("Jane", "Doe", "1975-01-01")
persons = [person1, person2]
for person in persons:  ①
    print(person.full_name())  ②
```

① We iterate over all the person objects in the list.
② In each iteration, the variable `person` refers to a different person.

Output

```
John Doe
Jane Doe
```

We will not create classes in this book, but we will use them. A lot indeed, as almost everything in Python is an object. So it is useful to have a basic understanding of what classes and objects are, and how they are used. For example, the `pandas` library uses classes to represent dataframes (`DataFrame`) and series (`Series`). There is a lot more to classes and objects in Python and in general. If you want to dive deeper, we recommend the Python documentation.[4]

2.9.5.2 Data Types and Their Declaration

In many other programming languages, variables must be declared to be of a specific type. This is not the case in Python. The type of a variable is determined by the value assigned to it. This is called *dynamic typing*. This is convenient (and powerful), but can also lead to errors that are not detected until the program is executed. For example, we can assign an integer value to a variable and then assign a string value to the same variable. If you are not aware of this change, you might get unexpected results.

Since version 3.5, Python allows to declare the type of a variable. This is called type hinting. Here is an example:

```
pi_value: float = 3.14  ①
message: str = "Python"
is_raining: bool = True
numbers: list[int] = []  ②
```

① This line declares that the variable `pi_value` is of type `float`. In a case like this it is not necessary to define the type explicitly, since the value already defines the type.
② This line declares that the variable `numbers` is a list of integers (`int`). In this case, it is necessary to define the type, since the value is an empty list and the type cannot be deduced from the value.

4 https://docs.python.org/3/tutorial/classes.html

Type hints can also be used to define the type of function parameters and return values. For details, see the documentation of the `typing` module in the standard library.

As mentioned, type hints are *hints* only, they are not enforced by the Python interpreter. We can still change the type of a variable later.

```
pi_value = "3.14"  # no error
```

However, there are external tools like *mypy* to identify cases like this and flag them as potential errors. Modern IDEs like *VSCode* are also able to check the code during development which makes adding type hints even more useful. Most of the widely used packages now support type hints.

2.10 Are We Sure We Made a Difference?

We found the average effect of our treatment was to reduce the number of hospital errors by almost one—0.92. However, we see that the variability from hospital to hospital is more than one. Some people define statistics as the art and science of finding patterns in the midst of chance variability. We think we found a difference, but could it be due to chance? There is enough variability that it is hard to be sure just by looking at the data. In Chapter 4 we will look at ways to assess whether the difference we see in this current example is real or might simply be the result of random variability in the numbers.

2.11 Is Chance Responsible? The Foundation of Hypothesis Testing

What do we mean when we say "by chance?" Let's illustrate with the simpler case of just one hospital that reduced its errors from seven in the first year to three in the second year under the no-fault program (this is the single hospital in Quebec we mentioned earlier).

2.11.1 Looking at Just One Hospital

For this single hospital case, we can formulate our "chance" question as follows:

If we have 10 errors and each error is assigned randomly to the non-treatment period or the treatment period (say, by coin flip), what is the probability that the treatment year could do as well as it actually did: a "reduction" of three errors, or even better?

A reduction from seven errors to three seems impressive (more than 50%), but a statistician recalling the vitamin E case might wonder if the change is real or if it could be just a fluke of chance.

A standard approach exists for answering the question "could chance be responsible?" This approach is called a *hypothesis* test. To conduct one, we first build a plausible mathematical model of what we mean by chance in the situation at hand. Then we use that model to estimate how likely it is, just by chance, to get a result as impressive as our actual result. If we find that an impressive improvement like the observed outcome would be very unlikely to happen by chance, we are inclined to reject chance as the explanation. If, on the other hand, our observed result seems quite possible according to our chance model, we conclude that chance is a reasonable explanation.

Why Perform Hypothesis Tests?

The logic of a hypothesis test seems convoluted, and is confusing. You have to make some hypothetical assumption ("chance is responsible"), and then try to disprove it. Why do this? Think of hypothesis tests as a tool to impose discipline on your brain, and avoid being led astray by chance. Engineers in industry conduct experiments and want to avoid embarking on costly process modifications suggested by chance results. Data scientists in marketing don't want to abandon tried and true marketing and branding campaigns on the basis of A/B tests that are statistical flukes. Drug developers are required to prove that the benefits of new drugs are real. The formal and mechanistic nature of hypothesis tests with their attendant technical terms adds to the confusion. Two groups in particular have formalized the process of assessing whether an observed result might be explainable by chance:

- Editors of the thousands of journals that report the results of scientific research, because they want to be sure that the results they publish are real and not chance occurrences, and
- Regulatory authorities, mainly in medicine, who want to be sure that the effects of drugs, treatments, etc., are real and are not due to chance.

Keeping in mind the "Is chance responsible" question, and what it means, can mitigate the confusion.

We will now conduct a hypothesis test for the Quebec hospital data. What do we mean by the outcome being "just" chance? What should that chance model look like? We mean that there's nothing remarkable going on—i.e. the no-fault reporting has no effect, and the $7 + 3 = 10$ major errors just happened to land seven in the first year and three in the second. If there is no treatment effect from no-fault reporting and only chance is operating, we might expect 50/50, or five in each year, but we would not *always* get five each year if the outcome were due to chance. One way we could see what might happen would be to just toss a coin 10 times, letting the 10 tosses represent the 10 major errors, and letting heads represent the first year and tails the second. Then a toss of `HTTHTTHHHH` would represent six in the first year and four in the second.

> **Try It Yourself**
> Toss a coin 10 times, and record the number of heads and the number of tails. We will call the 10 tosses one trial. Then repeat that trial 11 more times for a total of, say, 12 trials (120 tosses in all).
>
> Did you ever get seven (or more) heads in a trial of 10 tosses?
>
> At this stage, you have an initial impression of whether seven or more heads is a rare event. But you only did 12 trials. We picked 12 as an arbitrary number, just to get started. What's next?
>
> Let's try a Python script to toss the coins many more times.

Let's recap the building blocks of our model:

- A single coin flip, representing the allocation of a single error to this year (T in the above discussion) or the prior year (H in the above discussion);
- A series of 10 coin flips, representing a single simulation, also called a trial, that has the same *sample size* as the original sample of 10 errors;
- Twelve repetitions (an arbitrary number of times) of that simulation.

Another option is to sit down and figure out exactly what the probability is of getting seven heads, eight heads, nine heads, or 10 heads. You can see how this works for three tosses of a coin in Figure 2.2.

There are formulas that can be used to make these calculations, but for now, we will rely on simulations, especially their computer equivalents. Recall that our goal is to learn whether seven heads and only three tails is an extreme, i.e. unusual, occurrence. If we get lots of cases where we get eight heads, nine heads, etc., then clearly seven heads is not extreme or unusual.

Why do we count ≥ 7 instead of $= 7$? This is an important but often misunderstood point.

2.12 Probability

We have used the terms "probability" and "chance," and you probably have a good sense of what they mean. We hear talk of the "probability of precipitation" or

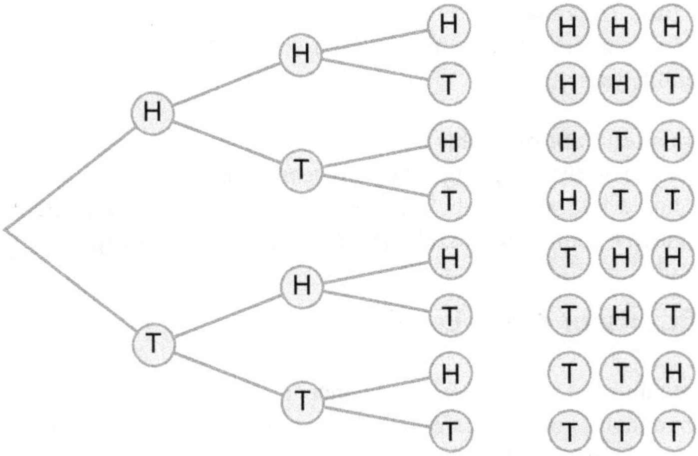

Figure 2.2 All possible outcomes for three coin tosses.

"chance of precipitation," and do not find it greatly puzzling. For now you can think of it terms of long term frequency: The probability of something happening is the proportion of time that it is expected to happen when the same process is repeated over and over. For example, if, over the years, we were to hear 100 weather forecasts of "20% chance of rain," we would expect to have rain 20 of those days. Probabilities are expressed as decimals ("0.20 probability of rain") or percentages. They must always lie between 0 (will never happen) and 1 (certain to happen), or 0% and 100%.

We will revisit probability in Chapter 5. For now, we return to the 10 coin tosses. It turns out that the probability of getting seven or more heads is about 0.1667. A bit later we'll get to the question of how this number is calculated or estimated (one way is by doing many more simulations), but for now let's continue with our story and see how we interpret this result.

2.12.1 Interpreting Our Result

The value of 0.1667 means such an outcome, i.e. seven or more heads, is not all that unusual, and the results from the single Quebec hospital might be due to chance.

Would you consider chance a reasonable explanation if there were 10 major errors the year before the change and none the year after? *Hint:* Use the coin tosses you already did.

Suppose it had turned out the other way? If our chance model had given a very low probability of the actual outcome, then we are inclined to reject chance as the main factor.

Definition: *P*-value If we examine the results of the chance model simulations, the probability of seeing a result as extreme as the observed value is called the *p-value* (or probability value).

Don't worry if this definition of *p*-value and the whole hypothesis testing process is not fully clear to you at this early stage. We will come back to it repeatedly.

The use of *p*-values is widespread; their use as *formal* decision-making criteria lies more in the *research* community than in the *data science* community.

2.13 Significance or Alpha Level

How unusual is too unusual to be ascribed to chance?

Most would agree that if you get an apparently "extreme" result 20% or 30% of the time under a chance model, that's not improbable enough to rule out chance as a possible cause.

Similarly, most would agree that if you get an extreme result only 1% of the time under a chance model, then we probably *can* rule out chance as a possible cause.

What about the territory in between? Custom and tradition typically set the threshold level of statistical significance as 5%. Given an apparently extreme result seen in real life, if the chance model produces such an "extreme" result less than 5% of the time, i.e. in 5% of the resampling trials, it is said to be statistically significant. This 5% level is from the writings, in the first part of the 20th century, of the great statistician, R. A. Fisher. He spoke of a frequency of one in 20 as being too rare to ascribe to chance.

This threshold level is termed "alpha" and is denoted by the Greek letter α. Alpha is determined before a study is done, usually by regulators or journal editors. As noted, it is typically set at 0.05.

Definition: Significance Level The significance level α is a threshold probability level set before doing a hypothesis test. If the hypothesis test yields a *p*-value at or below α the result is deemed statistically significant.

2.13.1 Increasing the Sample Size

Intuition tells us that small samples lead to fluke results. To see what happens when you increase the sample size, you could try doubling the number of coin-flips from 10 to 20, and observing the probability of getting an equivalent error reduction (from 14 to 6).

A comprehensive study like the one outlined earlier in this chapter provides for a much larger sample, and, importantly, allows the conclusions to be extended beyond a single hospital.

Law of Large Numbers

The impact of larger samples can be seen in "the law of large numbers." The law of large numbers states that, despite short-term average deviations from an event's theoretical mean, such as the chance of a coin landing heads, the long-run empirical–actual–average occurrence of the event will approach, with greater and greater precision, the theoretical mean. The short-run deviations get washed out in a flood of trials. During World War II, John Kerrich, a South African mathematician, was imprisoned in Denmark. In his idle moments, he conducted several probability experiments.

In one such experiment, he flipped a coin repeatedly, keeping track of the number of flips and the number of heads. After 20 flips, he was exactly even–10 heads and 10 tails. After 100 flips, he was down six heads–44 heads, 56 tails–or 6%. After 500 flips, he was up five heads–255 heads, 245 tails–or 1%. After 10,000 flips, he was up 67 heads or 0.67%.

A plot of all his results with the proportion of heads on the y-axis and the number of tosses on the x-axis shows a line that bounces around a lot on the left side, but settles down to a straighter and straighter line on the right side, tending towards 50% (see Figure 2.3a). A simulation of Kerrich's experiment (Figure 2.3b) shows that this behavior is not unusual.

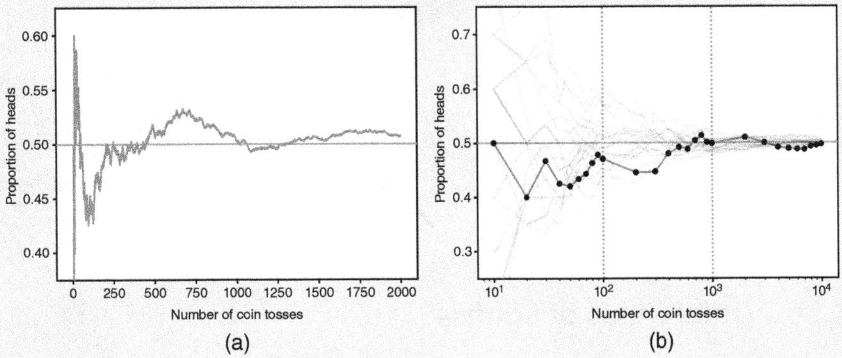

Figure 2.3 Kerrich coin tosses. Number of tosses on the x-axis and proportion of heads on the y-axis. (a) Original experiment data. (b) Simulation of up to 10,000 tosses; single result highlighted, additional repeats shown in grey.

Don't confuse the Law of Large Numbers with the popular conception of the Law of Averages.

Law of Large Numbers: Long run actual average will approach the theoretical average as the sample size grows.

Law of Averages: A vague term, sometimes meaning the above, but also used popularly to refer to the mistaken belief that, after a string of heads, the coin is "due" to land tails, thus preserving its 50/50 probability in the long run. One often encounters this concept in sports, e.g. a batter is "due" for a hit after a dry spell.

2.13.2 Simulating Probabilities with Random Numbers

We will be flipping coins or shuffling numbers mostly via random number equivalents generated by Python. For our purposes, we can think of a random number as the result of placing the digits 0 to 9 in a hat or box, shuffling the hat or box, and then drawing a digit. We modeled whether an error occurred under the treatment year or the pre-treatment year with a coin, and we could represent that by choosing either a "0" or a "1." Suppose a customer for a music streaming subscription service has a 15% probability of canceling the service in a given year. This could be modeled by generating a random integer between 1 and 100, and labeling 1–15 as "cancel" and 16–100 as "maintain subscription."

Most random numbers are produced by computer algorithms that produce a series of numbers that are effectively random and unpredictable, or at least sufficiently random for the purpose at hand. But the numbers are produced by an algorithm that is technically called a *pseudo-random number generator*. There have been many research studies and scholarly publications on the properties of random number generators (RNGs) and the computer algorithms they use to produce pseudo random numbers. Some are better than others; the details of how they work are beyond the scope of this text. We can simply think of RNGs as the computer equivalent of picking cards from a hat or box that has been well-shuffled.

2.14 Other Kinds of Studies

We have been dealing in this chapter mostly with experiments, and accounting for chance when we interpret the results. Experiments are costly, and most studies, published or otherwise, fall into two other categories.

- Sample surveys
- Observational studies

In **sample surveys**, we have a large group, called the population, about which we would like to know something. If we are able to obtain data on the entire population, then the study is called a census. In national election polls, the population might be all the people who will vote in the election.

Figure 2.4 The American Community Survey is a detailed version of the full census, based on a sample.

If we asked all of them to tell us how they would vote, we would have a census, but it would be difficult and expensive. So in many situations, we take a sample—some smaller part of the population. This allows us to spend time and effort selecting sample elements in a way (usually randomly) that well represents the population. The American Community Survey, for example, samples about 1% of the US population yet yields much more frequent, detailed, and accurate information than the 10-year Census (Figure 2.4). We will learn more about sample surveys in Chapter 7.

Observational studies work with pre-existing data. The study of nurses taking vitamin E is one example of an observational study. Among the most famous observational studies were the early investigations into the connection between smoking and lung cancer. This is a situation in which an experiment would not be ethical: we can not assign some people to smoke and others not to smoke if we believe smoking is harmful. Observational studies are vulnerable to "look-back bias," in which existing data is examined, sometimes extensively, in search of something interesting. We will learn more about this issue in Section 8.6. Since observational studies lack treatments and randomly selected controls, their conclusions have limitations.

One category of observational studies has grown rapidly to the point that it now accounts for most data analysis: statistical and machine learning predictive models, the foundation for AI. They can be considered observational studies since they rely upon existing data, not experiments.

Two practices can enhance confidence in observational studies:

1) Dividing the data randomly, and using one portion to derive a model and the other portion to test the application of the model.
2) Rather than using all the conveniently available data, use statistical sampling to get a fully representative sample of both conveniently available data and data that are not convenient.

2.15 When to Use Hypothesis Tests

Let's sum up when hypothesis tests are really used:

1) Experiments in business and industry (testing responses to web offers, prototyping a new manufacturing process, trying out a new chemical formulation, etc.). In such a case, the main point is to learn where the experiment results lie, compared to possible chance outcomes (i.e. a *p*-value). Setting alpha is not so important, since the decision-makers will want to judge the costs and benefits of implementation, further research, etc. on a case-by-case basis. The *p*-value is a factor, but not necessarily the deciding factor, in such analysis.
2) Studies for publication in scholarly or technical journals. The primary purpose here is to assure the readers that the effect reported is real, not due to chance, so alpha is typically set in advance (usually at 0.05) for a yes/no decision.
3) Studies for submission to governmental authorities (drug development, environmental compliance, legal cases). Drug development in particular has a long history of government control and a highly structured regulatory regime—an important goal is to assure the public that any claimed effects of drugs are real, and not the result of chance outcomes in studies.

2.16 Experiments Falling Short of the Gold Standard

We have described the attributes of a "gold standard" experiment:

- Group of subjects or cases, with a clearly defined intervention (treatment) that is relevant and important
- A clear and meaningful measure of the outcome
- Control group for comparison
- Random assignment
- Blinding of participants
- Blinding of experimenter and analyst
- Testing the result for statistical significance, to avoid being fooled by chance

Studies that meet *all* these criteria are the exception, rather than the rule. Medical studies come the closest, due to the implications for human safety and the need for regulatory approval. Most non-medical studies, however, work with existing data and lack treatments initiated by the person doing the study. Most data science analysis (predicting that an insurance claim is fraudulent, predicting that a web visitor will purchase, segmenting customers into groups, recommending products for purchase, and much more) falls in this category, involving large databases that are generated for routine business purposes. Marketing experiments with consumers can come close to the gold standard, since lists of prospective customers and web visitors can be randomly assigned to receive one message or another. Recipients of the message see what appears to be normal marketing material and are typically unaware they are part of an experiment.

The relative importance of formal hypothesis testing in the field of statistics has declined somewhat as companies steadily increase the amount of analysis they do with the vast quantities of data that they generate in the normal course of business. These data are not typically generated as part of a study, and their analysis often does not follow the protocol of an experiment or study that can be assessed with a standard hypothesis test. On the other hand, resampling methods (generating random data, shuffling existing data, sampling with replacement from existing data) are often used in less formal procedures to assess natural variability in data.

Try It Yourself

Think about the following scenarios, and how (and to what extent) the desirable attributes of an experiment might be met:

- A large consumer goods producer with both direct-to-consumer sales and sales via retail channels wants to test whether it should raise prices on a product line.
- A health organization wants to test the accuracy of image-processing algorithms in reading mammograms.
- A nonprofit wants to test different messaging in its membership renewal drive.
- Investors in a startup want the company to focus more on increasing profitability, and the company wants to know whether it can do this by laying off staff.
- An agricultural product company that supplies feed to farmers wants to test whether reducing the level of antibiotics in feed affects productivity.

2.17 Summary

Identifying truth and fact among a crowd of opinions, gut instincts, truisms, and assumptions is difficult. Just bringing data to the table is not sufficient, as the maxim about there being three types of lies, "lies, damned lies and statistics," illustrates.[5] Well-designed experiments can yield sound and reliable answers to questions. Key elements of these experiments include the use of a control group, random assignment, and blinding. Hypothesis testing is used to reduce the risk of

5 This phrase is often attributed to Mark Twain, though he credited Disraeli.

being fooled by chance, and resampling methods are an effective way of assessing chance variability. Relatively few studies can be designed to incorporate all attributes of a well-designed study, but it is still good to bear these principles in mind and adhere to as many as are practicable.

2.18 Python: Iterations and Conditional Execution

In Section 2.9, we discussed data structures in Python, including lists, tuples, dictionaries, and sets. Storing data is only the first step. To do something useful with data, we need to be able to combine and manipulate them. We may for example want to calculate the sum of all values in a list or determine the number of values above a given threshold. In this section, we will discuss how to iterate (repeatedly perform actions) over these data structures, and how to execute code conditionally.

2.18.1 `if` Statements

The `if` statement allows you to execute code conditionally, depending on whether a certain condition is met. It can be combined with the optional `elif` and `else` statements to execute code in different cases. Here is an example:

```
x = 123
if x < 0:  ①
    print('x is negative')
elif x == 0:  ②
    print('x is zero')
else: ③
    print('x is positive')
```

① The conditional execution begins with the `if` statement followed by a condition. The following code block is indented and executed only if the condition is `True`. In this case, the condition is `x < 0`, which is `False` because `x` is `123`. Therefore, the code block is not executed. Execution continues in the first line after the code block, which is the `elif` statement.

② The condition in the `elif` statement checks if `x` is equal to `0`. This condition is also `False`, so the next block is not excuted. Execution continues in the `else:` line.

③ The `else` statement does not have a condition. It is executed if none of the previous conditions were `True`. This is the case here, so the outcome of the whole `if-elif-else` statement is that "`x is positive` " is printed.

The `else` and `elif` statements are optional. The following example only has an `if` statement. After that, `x` is replaced with its absolute value.

```
x = -123
if x < 0:
    x = -x  ①
print(f'Absolute value of x: {x}')
```

① The code block is executed only if the x < 0, which is True in this case. In the if block, we make the negative number positive.

The elif statement is a shorthand for else: if. The initial example could also be written as:

```
x = 123
if x < 0:
    print('x is negative')
else:
    if x == 0:
        print('x is zero')
    else:
        print('x is positive')
```

If you have many conditions, the code will be indented more and more which will make it harder to understand the expressed logic.

2.18.2 for Statements

The for statement allows you to iterate over a sequence of values. The following example iterates over a list of numbers and prints each of them:

```
for x in [1, 2, 3, 4, 5]:  ①
    print(x)
```

① The statement is for *variable* in *iterable*. In this case, *iterable* is the list of numbers. Each of the elements in the list will be assigned to the variable x in sequence and the following code block executed. Like the if statement, the code block is indented. The code block is executed five times. In the first iteration, the value of x is 1, in the second iteration it is 2, and so on.

2.18.3 while Statements

The while statement allows you to execute a code block repeatedly as long as a condition is True. The following example prints the numbers from 1 to 5:

```
x = 1
while x <= 5:  ①
    print(x)
    x += 1  ②
```

① The statement is `while condition:`. The code block is executed as long as the condition is `True`. Here, the condition is `x <= 5`. In the first iteration, the value of `x` is `1`. At the end of the block, we increment `x` to `2`. The condition `x <= 5` is still `True`, so we start a second iteration with `x` equal to `2`, and so on. In the last iteration, the value of `x` is `6`, so the condition is `False` and the code block is not executed. In total, the code block executed five times.

② The statement `x += 1` is shorthand for `x = x + 1`. It increases the value of `x` by `1` in each iteration.

You need to be careful with `while` statements. If the condition is always `True`, the code block will be executed forever. This is called an *infinite loop*.

2.18.4 `break` and `continue` Statements

There will be cases when you want to execute a `for` or `while` loop, but skip some iterations. This can be done with the `continue` statement. The `continue` stops the execution of the remainder of the code block and continues with the next iteration. The following will print all odd numbers from 1 to 10:

```
for x in range(1, 11):
    if x % 2 == 0:  ①
        continue
    print(x)
```

① The `if` statement checks if the value of `x` is even using the modulo operator `%`. If this is the case, the `continue` statement is executed and the loop starts again with the next iteration.

Output

```
1
3
5
7
9
```

The `break` statement is similar to the `continue` statement. You use the `break` statement, if you want to stop a `for` or `while` loop before it has finished when a certain condition is met. This can be done with the `break` statement. The following example prints the numbers from 1 to 5, but stops when the value of `x` is `3`:

```
for x in [1, 2, 3, 4, 5]:
    if x == 3:  ①
        break
    print(x)
```

① The `if` statement checks if the value of `x` is `3`. If this is the case, the `break` statement is executed, which stops the `for` loop. The remainder of the code block is not executed for `x == 3`, so the number `3`, and the subsequent numbers, are not printed.

Output

```
1
2
```

2.18.5 Example: Calculate Mean, Standard Deviation, Subsetting

Let's see the control statements in action. As a first example, we calculate the mean of a list of numbers.

```
numbers = [12, 8, 9, 10, 11, 13, 9, 11, 10, 12]
sum_of_numbers = 0  ①
for x in numbers:  ②
    sum_of_numbers += x
mean = sum_of_numbers / len(numbers)  ③
print(f'Mean: {mean}')
```

① The variable `sum_of_numbers` is initialized with the value `0`. It will hold the sum of all numbers after the `for` loop.
② The `for` loop iterates over the numbers of the list and adds them to the variable `sum_of_numbers`.
③ After the loop, the variable `sum_of_numbers` contains the sum of all numbers. We divide this sum by the length of the list `numbers` to get the mean.

Next, we calculate the standard deviation of the list of numbers. The standard deviation is defined as the square root of the variance. The variance is the mean of the squared differences of the numbers from the mean. We will learn more about standard deviation and variance in Section 3.3.6. The following code calculates the variance and the standard deviation:

```
variance = 0
for x in numbers:
    variance += (x - mean) ** 2  ①
variance /= len(numbers)  ②
sd = variance ** 0.5  ③
print(f'Variance: {variance}')
print(f'Standard deviation: {sd}')
```

① The `for` loop iterates over all numbers of the list and adds the squared difference of each number from the mean to the variable `variance`.
② After the loop, the variable `variance` contains the sum of the squared differences of all numbers from the mean. We divide this sum by the length of the list `numbers` to get the variance.

③ The standard deviation is the square root of the variance. We calculate it by raising the variance to the power of `0.5`. We could also have used the *sqrt()* function from the `math` module.

Finally, we will use the `if` statement to subset the list of numbers. We will create a new list that contains only the numbers that are greater than the mean:

```
greater_than_mean = []  ①
for x in numbers:
    if x > mean:  ②
        greater_than_mean.append(x)  ③
print(f'Numbers greater than mean: {greater_than_mean}')
```

① We initialize an empty list `greater_than_mean` that will hold the numbers greater than the mean.
② The `for` loop iterates over all numbers of the list. The `if` statement checks if the number is greater than the mean. If this is the case, the number is appended to the list `greater_than_mean`.

2.18.6 List Comprehensions

In Python, simple `for` loops like the ones we've seen in the previous example can be rewritten using so-called *list comprehensions*. Consider these two examples:

```
result = []
for item in iterable:
    result.append(expression(item))
```

```
result = []
for item in iterable:
    if condition(item):
        result.append(expression(item))
```

The first example creates a list `result` by iterating over the elements of `iterable` and applying the function `expression` to each element. The second example creates a list `result` by iterating over the elements of `iterable` and applying the function `expression` to each element, but only if the condition `condition` is `True`. The two `for` loops can be rewritten using list comprehensions:

```
[expression(item) for item in iterable]
[expression(item) for item in iterable if condition(item)]
```

List comprehensions are a compact way to create lists and lead to more readable code.

Let's rewrite the previous examples using list comprehensions. Here is the calculation of the variance:

```
squared_differences = [(x - mean) ** 2 for x in numbers]  ①
variance = sum(squared_differences) / len(numbers)
print(f'Variance: {variance}, standard deviation: {variance ** 0.5}')  ②
```

① The expression here is `(x - mean) ** 2`.

② The f-string demonstrates how you can use an expression inside the curly brackets. The expression `variance ** 0.5` calculates the standard deviation temporarily and uses it in the f-string.

Instead of creating the intermediate array, we can also calculate the variance directly in a single step:

```
variance = sum((x - mean) ** 2 for x in numbers) / len(numbers)
```

This is not really a list comprehension. What we have here is a *generator expression*. It is similar to a list comprehension, but it does not create a list. Instead, it creates a generator object that can be used in a `for` loop or passed to a function like *sum()* or *list()*.[6]

The subsetting of the list of numbers can be rewritten like this:

```
greater_than_mean = [x for x in numbers if x > mean]
```

There are also dictionary comprehensions and set comprehensions. Here are two examples:

```
# set comprehension
numbers = [1, 2, 5, 1, 3, 2, 4, 5, 3, 4]
unique_squares = set()
for n in numbers:
    unique_squares.add(n**2)
unique_squares = {n**2 for n in numbers}  ①

# dictionary comprehension
pairs = [(1, 'one'), (2, 'two'), (3, 'three')]
number_to_word = {}
for key, value in pairs:
    number_to_word[key] = value
number_to_word = {key: value for key, value in pairs}  ②
```

① The `set` comprehension differs from list comprehensions by using curly braces instead of square brackets.

② The dictionary comprehension uses curly braces and a colon to separate the key and the value.

6 You can sometimes see the sum of a list calculated like this: `sum([x for x in numbers])`. This creates an intermediate list that is not needed. It is better to use `sum(x for x in numbers)`.

2.19 Python: Numpy, scipy, and pandas—The Workhorses of Data Science

You will often hear that Python is slow. Still, it is used in many data science applications that require fast processing of large amounts of data. How is this possible? The answer is that these applications use specialized Python packages that move data and time-intensive applications to highly optimized code that was written in a more suitable programming language. In this case, Python acts as the glue to manage the data and delegates the intensive processing to these packages. The most important packages for data science are `numpy`, `scipy`, and `pandas`.

2.19.1 Numpy

For example, `numpy` provides a highly efficient implementation of multidimensional numerical arrays. Let's look at an example and compare performance with a pure Python implementation. We will calculate the sum of all elements in a list of numbers. Let us first create two representations of a list of one million numbers:

```
import numpy as np  ①
numbers = list(range(1_000_000))  ②
numbers_np = np.arange(1_000_000)  ③
```

① The `import numpy as np` statement imports the `numpy` package and assigns it the alias `np`. This is a common convention that you will see everywhere. The `numpy` package is used so often, that it is easier to type `np` instead of `numpy`.

② The *range()* function creates an iterator that will create numbers from `0` to `999_999`. The *list()* function converts this to a list.

③ The *np.arange()* function creates a `numpy` array of numbers from `0` to `999_999`.

The `numbers` list is a pure Python list. The `numbers_np` array is a `numpy` array. The `numpy` array is more efficient in terms of memory usage and processing speed. It requires only about one-fifth of the memory required by the list. Let's see how the processing speed differs.

We can now calculate the sum of all numbers. First, we will use a `for` loop:

```
%%timeit -n 1 -r 5  ①
sum_of_numbers = 0
for x in numbers:
    sum_of_numbers += x
```

① The `%%timeit -n 1 -r 5` loop is a so-called *magic* command of the jupyter notebook. It will repeat the code five times and then outputs the average time.

Output
```
45.1 ms ± 3.09 ms per loop (mean ± std. dev. of 5 runs, 1 loop each)
```

It takes about 45 ms to calculate the sum of all numbers. This is not bad, but we can do better. If we replace the `for` loop with a call to the *sum()* function, the code requires only 6 ms.

```
%%timeit -n 1 -r 5
sum_of_numbers = sum(numbers)
```

Output
```
5.92 ms ± 349 µs per loop (mean ± std. dev. of 5 runs, 1 loop each)
```

Using the *sum* method of the `numpy` array, we can calculate the sum even faster in only 0.2 ms.

```
%%timeit -n 10 -r 5
sum_of_numbers = numbers_np.sum()
```

Output
```
184 µs ± 36.7 µs per loop (mean ± std. dev. of 5 runs, 10 loops
each)
```

This simple example shows that writing fast and efficient code in Python is possible if you use specialized packages. In fact, `numpy` is used by many other packages. It is at the core of `scipy`, `pandas`, and `scikit-learn`.

We already learned that `numpy` implements multidimensional numerical arrays and operations on these arrays. In addition, you can find functions for random number generation and linear algebra. We will encounter some of them in the following chapters. The `numpy.array` is at the core of this package. You can create a `numpy.array` from a list:

```
import numpy as np
numbers = [1, 2, 3, 4, 5]
x = np.array(numbers)  ①
print(x)
```

Output
```
[1 2 3 4 5]
```

There are many functions to manipulate and combine `numpy.arrays`. Here are a few examples.

```
import numpy as np
x = np.array([1, 2, 3, 4, 5])
y = np.array([6, 7, 8, 9, 10])
print(x + y)  ①
print(x * y)  ②
print(np.sqrt((x - 3) ** 2))  ③
print(x > 3)  ④
print(x[x > 3])  ⑤
```

① The + operator adds the elements of the two arrays. All operations act element-wise. This means the first element of the first array is added to the first element of the second array, the second element of the first array is added to the second element of the second array, and so on.
② The * operator multiplies the elements of the two arrays.
③ In this statement, 3 is first subtracted from each element in x. The ** operator then squares the elements of the resulting array. Finally, we take the square root of the array.
④ The > operator compares the elements of the array with 3. The result is a boolean array.
⑤ The [] operator selects the elements of the array that are True.

Output
```
[ 8 11 14 17 20]
[ 6 14 24 36 50]
[2. 1. 0. 1. 2.]
[False False False  True  True]
[4 5]
```

There are also a wide variety of functions to combine the elements of an array. For example,

```
import numpy as np
x = np.array([1, 2, 3, 4, 5])
print(x.sum())  ①
print(x.mean())  ②
print(x.cumsum())  ③
```

① The *sum* function calculates the sum of all elements in the array.
② The *mean* function calculates the mean of all elements in the array.
③ The *cumsum* function calculates the cumulative sum of all elements in the array. The first element of the resulting array is the first element of the original array, the second element is the sum of the first two elements of the original array, and so on.

Output
```
15
3.0
[ 1  3  6 10 15]
```
This short overview should allow you to follow the examples in this book. However, we strongly recommend to have a look at the numpy documentation (https://numpy.org) to learn more about the package.

2.19.2 scipy

The scipy provides many more functions for scientific computing. It includes functions for optimization, linear algebra, and statistics. We will encounter some of them in the following chapters and introduce them when needed. The scipy documentation (https://scipy.org) provides a good overview of the package.

2.19.3 Pandas

The name of the pandas package comes from ***panel data***. This is a reference to its use for the manipulation and analysis of dataframes. The pandas package provides two main data structures:

- Series for a sequence of data points. An example would be a time series of stock prices or the results of an experiment with a single variable.
- DataFrame for a table of data like we encountered in Section 2.7.1.

It also provides functions for reading and writing data from and to files, and for data manipulation.

We will use pandas extensively in the following chapters as a way of loading and collecting data and will highlight interesting features. For now, here is a short overview of its basic functionality. Review the pandas documentation (https://pandas.pydata.org) for detailed information.

2.19.3.1 Reading and Writing Data

Pandas has several functions for reading and writing data from and to files in different formats. The following example reads the data from the file data.csv and stores it in the variable df. The suffix .csv tells us that the file is in CSV format, this means fields are separated by a comma.

```
import pandas as pd  ①
df = pd.read_csv("hospitalerrors.csv")  ②
df.head()  ③
```

① The import pandas as pd statement imports the pandas package and assigns it to the common alias pd.

② The *read_csv* function reads a DataFrame from a CSV file. As given, the file must be in the same directory that the code is executed from. If the file is located in a different directory, you need to provide the full path to the file. The DataFrame is stored in the variable df.

③ The *head* method prints the first five rows of the DataFrame. This is useful to get a quick overview of the data. The output tells us that the data contains two columns, Control and Treatment. The rows underneath the column names are the first five rows of the data. The numbers 0 to 4 are the row indices. If not explicitly specified otherwise, pandas uses an index for the rows that starts at zero.

Output

```
   Control  Treatment
0        1          2
1        1          2
2        1          2
3        1          2
4        1          2
```

The *read_csv* function has many options to control how the data is read. For example, you can specify the column names or give the name of a column that should be used as an index. The `pandas` documentation contains a detailed description of all options.

You can also write a `DataFrame` to a file. The following example writes the `DataFrame` `df` to the file `data.csv`:

```
df.to_csv("data.csv", index=False)  ①
```

① The *to_csv* method writes the `DataFrame` to a CSV file. Without any options, the index is written as an additional unnamed column in the file. We use the `index=False` option to suppress this. If we don't set this option, we will need to deal with this additional column when reading the file.

2.19.3.2 Accessing Data

We now have a dataframe with the data from the file `hospitalerrors.csv`. We can access the data in different ways.

```
# Accessing a column
control = df["Control"]  ①
control = df.Control
# Accessing using row index and column names
control = df.loc[:, "Control"]  ②
row = df.loc[0]  # or df.loc[0, :]
values = df.loc[4:10, "Treatment"]
value = df.loc[0, "Control"]
# Accessing data using row and column numbers  ③
treatment = df.iloc[:, 1]
row = df.iloc[0, :]
value = df.iloc[10, 0]
```

① We've seen the `[]` operator as a way to access elements of lists and dictionaries. It has a similar function when used with a `pandas` `DataFrame` s. Directly applied to a `DataFrame`, it returns a single column as a `pandas` `Series` containing the data of the column when the column's name is used as the key. If a list of column names is used as the key, it returns a `DataFrame` containing the data of the columns matching the names in the list. As can be seen in the second example, you can also use the .operator to access a column. This will only work if the column name contains no spaces or special characters.

② The `.loc` attribute is another way to access the data. The first argument is the row index, the second argument is the column name. The `.loc` attribute can also be used to access a single value. The first argument is the row index, the second argument is the column name. Be careful if your row index is a number. The `.loc` attribute will interpret the number as a row index and not as the position in the dataframe. The `.loc` also understands slices as shown in the third example. Finally, if we refer to a specific row and column, the result is the value.

③ The `.iloc` attribute is similar to the `.loc` attribute, but uses row and column numbers instead of row and column names.

2.19.3.3 Manipulating Data

The `pandas` package provides a wide variety of functions to manipulate data. Here are a few examples:

```
# Adding or changing columns  ①
df["Constant value"] = 1
df["Sequence"] = range(len(df))
df["NewColumn"] = df["Control"] + df["Treatment"]
df["NewColumn"] = df["NewColumn"] * 2
# Removing a column  ②
df = df.drop(columns=["NewColumn"])
# Renaming columns  ③
df = df.rename(columns={"Control": "ControlGroup", "Treatment":
   "TreatmentGroup"})  ④
# Sorting the data
df = df.sort_values(by="ControlGroup")
```

① You create new columns by using the `[]` operator and assigning a value to the new column. The value can be a single value, a list, or a `numpy` array. In the first example, we add a new column with the name `"Constant value"` and the value `1` to the `DataFrame`. The second example adds a list. The third example combines two columns by adding the values elementwise. The fourth example takes an existing column, multiplies it by two, and replaces the original column with the new values.

② The *drop* method removes the specified columns from the `DataFrame` and returns a new dataframe.

③ The *rename* method renames the specified columns. The `columns` argument is a dictionary that maps the old column names to the new column names. The *rename* method returns a new `DataFrame` with the renamed columns. If you want to modify the original `DataFrame`, you need to assign the result of the *rename* method to the original `DataFrame`.

④ The *sort_values* method sorts the `DataFrame` by the specified column. The `by` argument is the name of the column to sort by. The *sort_values* method returns a new `DataFrame` with the rows sorted by the specified column. If you want to modify the original `DataFrame`, you need to assign the result of the *sort_values* method to the original `DataFrame`.

Even though it can look like we copy data when using `pandas`, most of the time changes are done with minimal changes to the underlying data structure. This makes it very efficient in handling data. Should you need to create an independent copy of the data, use the *df.copy()* method.

2.19.3.4 Iterating Over a `DataFrame`

You can iterate over the rows of a `DataFrame` using the *iterrows* method. The following example prints the index and the `ControlGroup` information for the first two rows of the `DataFrame` `df`:

```
for index, row in df.iterrows():  ①
    if index > 1:
        break
    print(f'Index: {index}')
    print(row)  ②
```

① The *iterrows* method returns a pair of the row index and the row in each iteration.

② The `row` is a `Series` containing the values of the row. You can access values in each row using the column names.

Output

```
Index: 0
ControlGroup      1
TreatmentGroup    2
Constant value    1
Sequence          0
Name: 0, dtype: int64
Index: 1
ControlGroup      1
TreatmentGroup    2
Constant value    1
Sequence          1
Name: 1, dtype: int64
```

2.19.3.5 And a Lot More

`Pandas` has a lot more to offer. We will introduce additional functionality when needed. For now, we recommend to have a look at the `pandas` documentation (https://pandas.pydata.org) to learn more about the package.

Exercises

2.1 Dart throws that miss the target (extreme ones miss the dartboard entirely and nick the wall) are analogous to errors in statistical estimates. Do the nicks in the wall in Figure 2.5 of a dartboard exhibit bias? If so, what do you think causes it?

Figure 2.5 Dart throw misses.
Source: Peter Bruce (Book Author).

2.2 Answer the following questions, referring to the *pulse.csv* data. You do not need to execute any simulations, just describe them as a series of steps that you might take with cards and a hat, or with random numbers.
Hint: How many cards would you need, and how would you shuffle and deal them out into samples?

a) Is the proportion who ran different for males vs. females? Describe how you might use cards marked "1" for "ran" and "0" for "didn't run" in a resampling experiment to test whether chance might be responsible for any difference.
b) Does the "before" pulse rate differ between smokers and nonsmokers? Medical theory suggests that smoking elevates the pulse rate. Describe how you might use cards marked with pulse rates in a resampling experiment to test whether chance might be responsible for any difference.
c) We would expect running in place to affect a person's pulse rate. Does the "before" pulse rate differ from the "after" pulse rate? Describe how you might use cards marked with pulse rates in a resampling experiment to test whether chance might be responsible for any difference.

2.3 Which is a more effective marketing message—a plain text email, or an email with pictures and design elements? Marketing professionals typically recommend the latter. Statistics.com performed an A/B test, sending an email to a list that was randomly split into two groups. (The email service did not produce an even split, but that did not mean it was not random. Random does not mean 50/50—think of throwing a die: if the die lands "1" the text message is sent, otherwise the email with images is sent.) Here are the results:

- Group A, plain text: 71 sent, 13 opens
- Group B, same content with images: 355 sent, 47 opens

a) The marketing manager worries that the imbalance in the sample sizes renders the experiment invalid. He thinks that, to perform a valid test of statistical significance, the two groups must be about the same size. Is this correct?

b) Looking at the results, how do groups A and B differ? Express the answer in units that are relevant for comparison. Is this what you were expecting?

c) Describe how you might use cards marked with 0's and 1's in a resampling experiment to test whether chance might be responsible for any difference.

2.4 DISCUSS A common problem in the TV streaming business is that customers sign up to watch just a few series, then cancel a couple of months later, after they are finished watching the shows they are interested in. This problem is particularly acute when show episodes are released all at once for "binge-watching," which is thought to generate maximum impact and generate word-of-mouth publicity. Discuss whether and how an experiment might play a role in determining whether it is better for the broadcaster to release episodes all at once or space them out over time.

2.5 Consumer packaged goods (CPG) companies sell some goods online but generate most of their revenue from sales through brick and mortar retail companies. A CPG company typically manages a number of brands, each of which has multiple products. For example, Unilever, a large CPG based in the United Kingdom, owns over 400 brands. Just one of its brands, Dove, sells body washes, hand, and body lotions, facial cleansers, deodorants, shampoos, conditioners, and hair styling products. Each product is sold in multiple sizes and variations (e.g. different soap scents), so the number of individual "stockkeeping units" (SKU's) on offer from a CPG might be in the tens of thousands. A typical grocery or drug store has room to stock only a small fraction of all the products on offer from CPGs. A major challenge for the CPG is obtaining and retaining "shelf space" at major retailers. CPG companies promote their products through advertising, issuance of discount coupons to consumers, and "trade discounts" offered to retailers.

Trade discounts are product-specific discounts offered to individual retailers that can fund co-advertising, or simply serve as an incentive to the retailer to promote the discounted product. The company is debating whether to reduce trade discounts and boost coupons, as is done in some of its Latin American markets. Discuss the possible design of an experiment to shed light on whether this is a good idea.

2.6 A company reported monthly sales figures for a retail store over the past year. The sales figures are as follows (Jan. to Dec., $ million):

```
sales_figures = [6.5, 7.5, 9.3, 10.2, 11, 9.5, 10.5,
                 12.5, 13, 14, 15.5, 16]
```

Analyze the sales figures using Python to answer the following questions:

a) What is the average monthly sales figure?
b) In which months were the sales figures above the average?
c) What is the maximum sales figure and in which month did it occur?
d) Print the sales figures, skipping any month where sales were below 10.
e) Print the sales figures for each month, stopping as soon as a month has lower sales than the previous month.
f) *Optional:* Try to combine some of the steps or calculations to optimize your code, reducing redundancy.

2.7 In this problem, we will get experience with writing `for` loops and `if` statements in Python. For each of the following exercises, write first a solution using `for` loop and `if` statements and then write a second solution that uses list comprehensions.

a) Create a list which contains the numbers 1 to 10. Then using a `for` loop, create a new list that contains the squares of each number. Use a list comprehension to achieve the same result.
b) Create a list which contains a mix of numbers between −5 and 5. Then using a `for` loop, create a new list that contains only the positive numbers. Use a list comprehension to achieve the same result.

2.8 Use `pandas` to read the *pulse.csv* data and answer the following questions:

a) What columns are in the dataset?
b) What are the average pulse rates before and after the exercise? By how much did the pulse rate change?
c) How many smokers and non-smokers were studied?
d) On average, how much did the pulse rate change for smokers and non-smokers?

3

Exploring and Displaying the Data

Many data analysis mistakes can be avoided by first looking at summaries and visualizations of the data. After completing this chapter you should be able to:

- Calculate measures of central location, such as mean and median
- Judge which measure of central location is appropriate for a particular scenario
- Calculate measures of variation, such as variance and percentiles
- Measure distance between records
- Produce a frequency table
- Interpret a box plot and histogram
- Describe what an outlier is

3.1 Exploratory Data Analysis

Some data analyses begin without a preconceived hypothesis. Space scientists want to examine samples brought back from the moon to see what elements are present. Marketers want to know about the characteristics of people who buy a given product. Business researchers want to know about the financial management structures of successful firms.

In other cases, a hypothesis is formed prior to the collection of data. Market researchers may want to test a theory that urban residents are more likely to purchase a certain product than rural residents. In the case study we have been looking at, hospital administrators want to test the proposition that no-fault reporting for errors will reduce the number of major medical errors.

In either case, it is good to conduct exploratory data analysis (EDA) to summarize and display the data to develop greater understanding. There is one important distinction:

Statistics for Data Science and Analytics, First Edition. Peter C. Bruce, Peter Gedeck, and Janet Dobbins.

Companion website: www.wiley.com/go/Wiley_Statistics_for_Data

- If a hypothesis is developed out of the data exploration, it should be tested for statistical significance with *new* data. If you test with the same data that helped suggest the hypothesis, there will be a bias in favor of finding that the result is significant.

3.2 What to Measure—Central Location

Part of the plan for any experiment will be the choice of what to measure to see if the treatment works. In our hospital experiment, we chose to measure "average (mean) reduction in errors." This is a good place to review the standard measures with which statisticians are concerned: central location of, and variation in, the data.

3.2.1 Mean

The *mean* is the average value—the sum of all the values divided by the number of values. It is generally what we use unless we have some reason not to use it.

Consider the following set of numbers: $\{3\ 5\ 1\ 2\}$

The mean is $\dfrac{3+5+1+2}{4} = \dfrac{11}{4} = 2.75$.

You will encounter the following symbols for the mean:

- $\bar{x}$ represents the mean of a *sample* from a population. It is written as x-bar in inline text.
- μ represents the mean of a *population*. The symbol is the Greek letter mu.

Why make the distinction? Information about samples is observed, and information about large populations is often inferred from smaller samples. Statisticians like to keep the two things separate in the symbology.

3.2.2 Median

The *median* is the middle number on a ranked list of numbers. Table 3.1 shows the ranked data for both groups of hospitals.

The middle number on each list would be the 13th value (leaving 12 numbers above and 12 below). If there is an even number of data values, the middle value is the average of the two values that divide the sorted data into upper and lower halves.

We find that the median is the same for both lists. It's two. This is not unusual for data with a lot of repeated values. The median is a blunt instrument for describing such data. From what we have seen so far, the groups seem to be different.

Table 3.1 Hospital error reductions, treatment, and control groups.

Control	Treatment
1	2
1	2
1	2
1	2
1	2
1	2
1	2
1	2
1	2
1	2
1	2
1	2
2	2
2	2
2	2
2	2
2	2
2	2
2	3
2	3
3	4
3	4
4	5
4	6
5	9

The median does not capture that. Looking at the numbers, you can see the problem. In the control group, the numbers coming before the two at Position 13 are all ones; for the treatment group they are all twos. The median reflects what is happening at the center of the sorted data, but not what is happening above or below the center.

The median is more typically used for data measured over a broad range where we want to get an idea of the typical case without letting extreme cases skew the results. Let's say we want to look at typical household wealth in neighborhoods around Lake Washington in Seattle. In comparing the Medina neighborhood to the Windermere neighborhood, using the mean would produce very different results because Bill Gates lives in Medina. If we use the median, it won't matter how rich Bill Gates is—the position of the middle observation will remain the same.

A student gave seven as the median of the numbers 3,9,7,4,5. What do you think they did wrong?

3.2.3 Mode

The *mode* is the value that appears most often in the data, assuming there is such a value. In most parts of the United States, the mode for religious preference would be Christian. For our data on errors, the mode is two for all 50 subjects and one for the control group. The mode is the only simple summary statistic for categorical data, and it is widely used for that. At different times in the history of the United States, the mode for the make of new cars sold each year has been Buick, Ford, Chevrolet, and Toyota. The mode is rarely used for measurement data.

3.2.4 Expected Value

The expected value is calculated as follows.

1) Multiply each outcome by its probability of occurring.
2) Sum these values.

For example, suppose a local charitable organization organizes a game in which contestants purchase the right to spin a giant wheel with 50 equal sized sections, and an indicator that points to a section when the wheel stops spinning. The right to spin the wheel costs \$5 per spin. One section is marked \$50—that's how much the purchaser wins if the spinner ends up on that section. Five sections are marked \$15, 10 sections are marked \$5, and the remaining sections are marked \$0.

To calculate the expected value of a spin, the outcomes, with the purchase price of the spin subtracted from the prize, are multiplied by their probabilities and then summed.

$$\text{EV} = \frac{1}{50}(\$50 - \$5) + \frac{5}{50}(\$15 - \$5) + \frac{10}{50}(\$5 - \$5) + \frac{34}{50}(\$0 - \$5)$$

$$\text{EV} = -\$1.50$$

The expected value of a ticket is negative for the purchaser (which produces profit for the charitable organization). For each ticket you purchase, you can expect to lose, on average, approximately \$1.50. Of course, you will not lose exactly \$1.50

in any of the above scenarios. Rather, the $1.50 is what you would lose per ticket, on average, if you kept playing this game indefinitely.

The expected value is really a fancier mean: it adds the ideas of future expectations and probability weights. Expected value is a fundamental concept in business valuation and capital budgeting—the expected number of barrels of oil a new well might produce, for example, the expected value of five-years of profits from a new acquisition, or the expected cost savings from a new patient management software at a clinic.

3.2.5 Proportions for Binary Data

When you have binary data, such as approve/disapprove, survive/die, the measure of central tendency is the proportion. An example would be the proportion of web visitors who click on a specified link. The proportion for binary data fully defines the data—once you know the proportion, you know all the values. For example, if you have a sample of 50 zeros and ones, and the proportion for one is 60%, then you know that there are 30 ones and 20 zeros.

3.2.5.1 Percents

Percents are simply proportions multiplied by 100. Percents are often used in reporting, since they can be understood and visualized more easily and intuitively than proportions.

3.3 What to Measure–Variability

If all the hospitals in the control group had one fewer error and all those in the treatment group had two fewer, our job would be easy. We would be very confident that the treatment improved the reduction in the number of errors by exactly one. Instead, we have a lot of variability in both batches of numbers.

Variability lies at the heart of statistics: measuring it, reducing it, distinguishing random from "real" variability, identifying the various sources of real variability, and making decisions in the presence of it.

Just as there are different ways to measure central tendency—mean, median, mode—there are also different ways to measure variability.

3.3.1 Range

The *range* of a batch of numbers is the difference between the largest and smallest number. Referring to Table 3.1, the range for the control group is $5 - 1 = 4$. Note that in statistics the range is a single number.

Try It Yourself
Referring to the same table, what is the range for the treatment group?

The range is very sensitive to outliers. Recall the two similar Seattle neighborhoods—Windermere and Medina. The range of income in Medina, where Bill Gates lives, will be much larger than the range in Windermere.

3.3.2 Percentiles

One way to get around the sensitivity of the range to outliers is to go in a bit from each end and take the difference from there. For example, we could take the range between the 10th percentile and the 90th percentile. This would eliminate the influence of extreme observations.

Definition: *P*th Percentile In a population or a sample, the *P*th percentile is a value such that at least *P* percent of the values take on this value or less and at least (100 − *P*) percent of the values take on this value or more. Sometimes there is a single value in the data that satisfies this requirement, and sometimes there are two. In the latter case, it is best to take the midpoint between the two values that do. Software may have slightly differing approaches that can produce differing answers.

More intuitively: to find the 80th percentile, sort the data. Then, starting with the smallest value, proceed 80% of the way to the largest value. Percentiles are encountered all the time—such as your score on a test relative to a large class, or household income level thresholds ("above the top one percent").

3.3.3 Interquartile Range

One common measure of variation is to take the difference between the 25th percentile and the 75th percentile.

Definition: Interquartile Range The interquartile range (or IQR) is the 75th percentile value minus the 25th percentile value. The 25th percentile is the first quartile, the 50th percentile is the second quartile, also called the median, and the 75th percentile is the third quartile. The 25th and 75th percentiles are also called *hinges*.

Here is a simple example: 3, 1, 5, 3, 6, 7, 2, 9. We sort these to get 1, 2, 3, 3, 5, 6, 7, 9. The 25th percentile is at 2.5 and the 75th percentile is at 6.5, so the IQR is 6.5–2.5 = 4. Again, software can have slightly differing approaches that yield different answers.

> **Try It Yourself**
> Find the IQR for the control data, the treatment data, and for all 50 observations combined.

3.3.4 Deviations and Residuals

There are also a number of measures of variability based on deviations from some typical value. Such deviations are called residuals.

Definition: Residual A residual is a difference between a mean value and an observed value or the difference between a value predicted by a statistical model and an actual observed value.

For the numbers $1, 4, 4$ the mean is 3 and the median is 4. The deviations from the mean are the differences

$$1 - 3 = -2$$
$$4 - 3 = 1$$
$$4 - 3 = 1$$

3.3.5 Mean Absolute Deviation

One way to measure variability is to take some kind of typical value for these residuals. We could take the absolute values of the deviations—$\{2, 1, 1\}$ in the above case—and then average them: $(2 + 1 + 1)/3 = 1.33$. Taking the deviations themselves without taking the absolute values would not tell us much—the negative deviations offset the positive ones. This always happens with the mean.

3.3.6 Variance and Standard Deviation

Another way to deal with the problem of positive residuals offsetting negative ones is by squaring the residuals.

Definition: Variance The *variance* is the mean of the squared residuals, where μ = population mean, x represents the individual population values, and n = number of observations in the data.

$$\text{Variance} = \sigma^2 = \frac{1}{n} \sum (x - \mu)^2$$

Definition: Standard Deviation The *standard deviation* σ is the square root of the variance. The symbol σ is the Greek letter *sigma* and commonly denotes the standard deviation.

The standard deviation is a fairly universal measure of variability in statistics for two reasons:

- It measures typical variation in the same units and scale as the original data and
- It is mathematically convenient, as squares and square roots can effectively be plugged into more complex formulas.

Absolute values encounter problems on the latter front, though some of those have been overcome as computational elegance fades in importance as computing power grows.

> **Try It Yourself**
> Find the variance and standard deviation of 8, 1, 4, 2, 5 by hand. Is the standard deviation in the ballpark of the residuals, i.e. the same order of magnitude?

3.3.6.1 Denominator of N or $N-1$?

Often, when considering "what to measure," we are interested not just in the dataset at hand, but a larger population from which it came. Intuitively, we are tempted to estimate a population metric by using the same metric in the sample. Most sample measures of *location* (mean, median) are *unbiased* estimators of the population value. Measures of *variability*, however, are *biased.*

Consider the range of the above values in the "Try it Yourself." Would that be an accurate estimate of the range of the entire population from which that sample came? Probably not—it is likely an under-estimate of the range, since there are bound to be more extreme values in a larger population. The same is true of the variance and standard deviation. If you use the intuitive denominator of N, you will underestimate the true value in the population.

However, if you divide by $N - 1$ instead of N, the variance and standard deviation from a sample become unbiased estimators of the population values. A mathematical proof is beyond our scope here, but you can confirm it with a resampling simulation in which you repeatedly take small samples from a known larger population. Calculate the variance for each resample using both $N - 1$ and N, and compare the average of those resample variances to the variance of the known population they were drawn from.

> **Try It Yourself**
> This exercise illustrates how the use of N in the denominator for calculating sample variance leads to a biased result, relative to the population variance. In Python, randomly generate a population of 1000 values. It doesn't matter what population you generate—let's say a population of 1000 randomly

selected numbers between 0 and 9. In Python, you can do this with the command `random.choices(range(10), k=1000)`. Next, find the variance of this population using N in the denominator. Then repeatedly take resamples of size 10, and calculate the variance for each resample according to the same population formula. How does the mean of the resample variances compare to the population variance? Now repeat the exercise using $N - 1$ in the denominator. How does the mean of the resample variances compare to the population variance? *Hint:* See Problem 4.12. for a solution.

3.3.7 Population Variance

If the observations at hand constitute the entire population that you are interested in, you can, in fact, use N as the denominator in the variance calculation. If N is large, though, it won't make much difference whether you divide by N or $N - 1$.

3.3.8 Degrees of Freedom

In statistics you will encounter the term *degrees of freedom*. Its exact definition is not needed here, but the concept can be illustrated. Let's say you have three observations and you know that their variance is x. Once you know the first two values, the third is predetermined by the first two and the value for the variance. We say there are $n - 1$, in this case, two, degrees of freedom. The denominator in the sample variance formula is the number of degrees of freedom.

3.4 What to Measure–Distance (Nearness)

The concept of *distance* is of particular interest to the *data science* community identified at the beginning of the Introduction.

Consider a poll in which respondents are asked to assess their preferences for the musical genres listed in Table 3.2. Ratings are on a scale of 1 (dislike) to 10 (like), and we have the poll results from three students.

Consider person C. Is she more like person A or person B? Looking at the scores, our guess would be that person C is more like person A. We can measure this distance statistically by subtracting one vector from another, squaring the differences so they are all positive, summing them so we have a single number, then taking the square root so the original scale is restored. If the two records are identical, the distance between them will be 0.

Table 3.2 Musical genre preferences.

Person	Rock	Hip-Hop	Country	Jazz	New Age
A	7	1	9	1	3
B	4	9	1	3	1
C	9	1	7	2	2

Definition: Vector A vector is a row of numbers. Vector arithmetic is done by performing the operation on the corresponding elements of each vector, resulting in a new vector of sums, differences, products, etc.

The statistic that we have described is "Euclidean Distance." Here is the formula, followed by the calculations.

As a general example, assume we have two vectors, w and x, each containing n values. The Euclidean Distance between the two vectors is

$$\text{Euclidean Distance} = \sqrt{(w_1 - x_1)^2 + (w_2 - x_2)^2 + \cdots (w_n - x_n)^2}$$

If you look carefully at the formula, you might recognize that this is the multidimensional version of the formula for the distance between two points that you may have learned in geometry.

For a specific example, the Euclidean Distance between vectors A and B—a measure of how alike person A is to person B—in Table 3.2 is calculated in Table 3.3.

In Table 3.3, the sum of the squares of the differences of each row is 145. The square root of 145 is 12.04, which is the Euclidean Distance between the two vectors representing person A and person B. Looking back at the data in Table 3.2, try the following problem:

? Let's say that you run a digital music service. A, B, and C are all customers of yours, and A and B have both just made downloads. You want to recommend one of these downloads for C. Which one would you recommend?

Table 3.3 Musical genre preferences.

	Rock	Hip-Hop	Country	Jazz	New Age
Person A	7	1	9	1	3
Person B	4	9	1	3	1
$(A - B)^2$	9	64	64	4	4
$\sum (A - B)^2$	145				
Euclidean Distance	12.04				

In this small-scale problem you can guess the answer just by looking at the data, but a practical business solution requires automation, which needs a distance metric that can be calculated. Distance measures are used in statistics for multiple purposes:

- Finding clusters or segments of customers who are like one another.
- Classifying records by assigning them the same class as nearby records.
- Locating outliers, e.g. airport security screening.
- Finding the distance to a benchmark. For example, if you have a list of symptoms for an individual, what disease is it closest to?

3.5 Test Statistic

Let's continue with our analysis of the hospital data, using the means for reduction in major errors. The treatment seems to reduce the number of errors by 2.80–1.88, or nearly one. But there are other ways to look at this. The ratio $2.80/1.88 = 1.49$ gives another comparison. It says the reduction in errors for the treatment group is 1.49 times that in the control group or nearly 50% greater.[1]

Our *test statistic* will be calculated as follows.

1) Measure the number of major medical errors for each hospital for the year before and the year after the treatment is initiated, and find the reduction: errors before minus errors after.
2) Calculate the mean reduction for the control group and the treatment group.
3) Find the difference: treatment minus control = 0.92.

Important: Throughout this example, we will be talking about "reductions in number of errors," not in the number of errors.

A test statistic is the key measurement that we use to judge the results of the experiment or study. It is best to select the test statistic beforehand, to avoid the temptation to "shop around" for different statistics after the results are in, looking for the most interesting result.

3.5.1 Test Statistic for this Study

The test statistic for this study is "mean reduction in errors (treatment) minus mean reduction in errors (control)."

---

1 Just counting errors treats them all alike. Ideally, we'd like to have some sort of measurement that accounts for the severity for each error. This introduces complexity into both the implementation of the study and the analysis.

3.6 Examining and Displaying the Data

The difference in error-reduction means between the two hospital groups is the best single metric to answer our original question, but it is not the whole story. We'd also like to learn more about the variability in the data.

3.6.1 Frequency Tables

Now that we have our results on 50 individual hospitals, we need a way to summarize and compare the treatments and controls as groups. We will first look at summaries that are numbers, then summaries that are pictures. One numerical summary we could make here is a table of values and how often those values occur, i.e. their frequencies (see Table 3.4).

This is called a *frequency table* or frequency distribution. Let's interpret a couple of rows.

- The first row tells us that 12 of the 25 control hospitals had a reduction in errors of one, and that none of the treatment group hospitals had a reduction in errors of one.
- The second row tells us that eight of the 25 control hospitals had a reduction in errors of two, and that 18 of the treatment group hospitals had a reduction in errors of two, so that a total of 26 hospitals had a reduction in errors of two.
- The bottom row is a summary: it tells us that there were 25 control group hospitals and 25 treatment group hospitals, for a total of 50 hospitals.

Notice that if we did not have a control group, we would overestimate the success of the treatment. All the hospitals improved, even in the control group. Still, it looks like the treatment group showed more improvement.

Table 3.4 Frequency distribution—reduction in errors.

Value	Control	Treatment	Total
1	12	0	12
2	8	18	26
3	2	2	4
4	2	2	4
5	1	1	2
6	0	1	1
9	0	1	1
All	25	25	50

Table 3.5 Error reduction frequency table (control).

Error reduction	Freq.	Cumulative Frequency	Relative Frequency
1	12	12	0.48
2	8	20	0.32
3	2	22	0.08
4	2	24	0.08
5	1	25	0.04
6	0	25	0.00
9	0	25	0.00
All	25	25	1.00

Frequency tables often include *cumulative or relative frequencies*. Row two in Table 3.5 says that eight hospitals had a reduction in errors of two, while a reduction of two *or fewer* errors showed up 20 times, and the eight times out of 25 constituted $0.32 = 32\%$ of the total.

Though it is not shown above, we could also calculate *cumulative relative frequency* in the same way that cumulative frequency is calculated—by adding together the current row with the preceding rows. For example, the cumulative relative frequency for the third row (error reduction = 3) is $0.48 + 0.32 + 0.08 = 0.88$.

> **Try It Yourself**
> Compute cumulative frequencies and relative frequencies for the treatment hospitals. Can you also find cumulative relative frequencies?

3.6.2 Histograms

Let's turn the frequency table into a plot—a frequency histogram. Figure 3.1 shows a histogram of the error reductions for the treatment group.

Interpreting Figure 3.1, we see that 18 hospitals reduced the number of errors by two, two hospitals reduced the number of errors by three, and so on. No hospital had an error reduction of seven or eight, but the histogram must leave room for these values to present an accurate picture.

In the above figure, the histogram is relatively easy to make—there are only eight possible values, so we can have a vertical bar for each value.

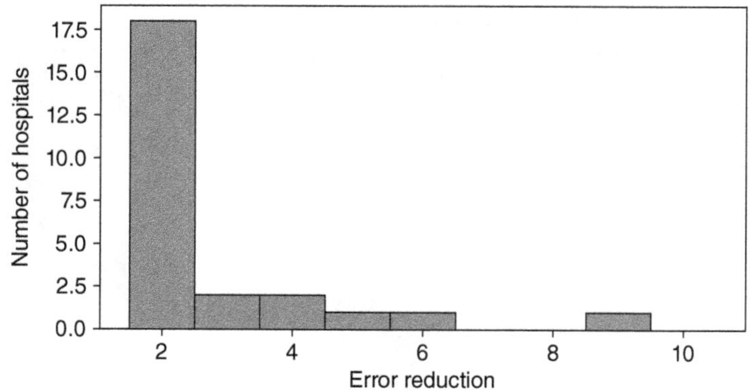

Figure 3.1 Frequency histogram of error reductions in treatment group.

If we are plotting more complex data—say, hospital sizes—we will not have enough room or visibility to devote one bar to each value. Instead, we group the data into bins. It is important that the bins be (1) equally-sized and (2) *contiguous*. By *contiguous*, we mean that the data range is divided up into equally sized bins, even if some bins have no data, like 7 and 8 above. Consider Figure 3.2, which shows hypothetical data for hospital sizes in a mid-sized state.

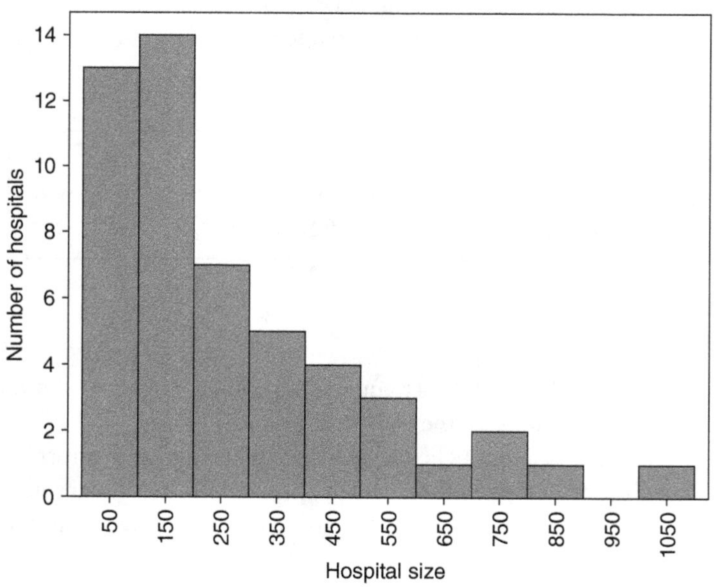

Figure 3.2 Hospital sizes by number of beds (hypothetical data for a mid-sized state).

Control	Errors	Treatment
XXXXXXXXXXXX	1	
XXXXXXXX	2	XXXXXXXXXXXXXXXXXX
XX	3	XX
XX	4	XX
X	5	X
	6	X
	7	
	8	
	9	X

Figure 3.3 Back-to-back histogram.

We can see that there were 13 hospitals, with 0 to 99 beds. There were 14 hospitals with 100 to 199 beds, etc.

Deciding how to display these data is not a trivial matter for a computer. The program must decide where the bin boundaries are and messiness can arise. Often, the algorithm that is used results in non-integer values on the x-axis, which may not make sense.

Figure 3.3 shows a *back-to-back* horizontal histogram of error reductions showing the control and treatment groups separately on either side of the Errors column.

3.6.3 Bar Chart

The histogram looks similar to another chart that is widely used—the bar chart. Figure 3.4 is a bar chart that depicts the number of applicants to different departments in the UC Berkeley graduate school in the fall of 1973 (they are discussed in Freedman, Pisani, Purves and Adhikari, *Statistics*, W. W. Norton, 4th ed, 1991):

Bar charts differ from histograms in two respects :

- unlike histograms, bar charts are usually drawn with gaps between the bars.
- unlike histograms, where the x-axis shows a progression of values on a continuous scale, the x-axis values on bar charts represent categories that are separate from one another and not part of a numerical continuum.

3.6.4 Box Plots

Let's look at another plot of the data distribution. The *boxplot* shows key percentiles of the distribution, as well as outliers.

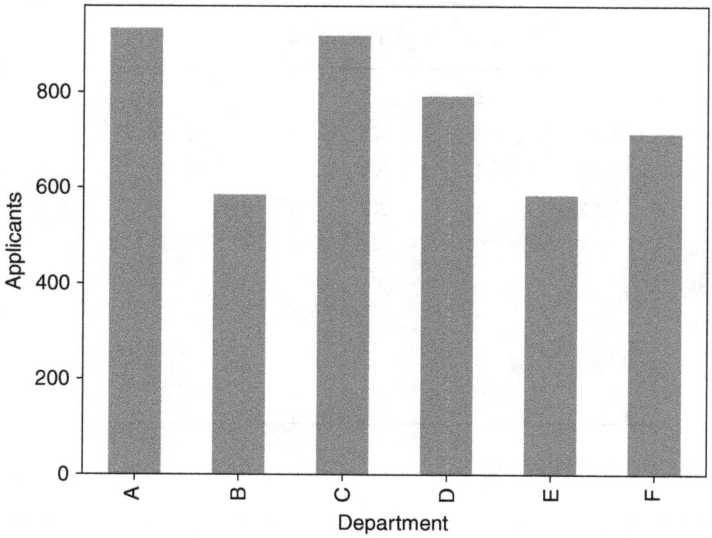

Figure 3.4 Bar chart—applicants by department.

- A central box encloses the central half of the data—the top of the box is at the 75th percentile, and the bottom of the box is at the 25th percentile.
- The median is marked with a line.
- "Whiskers" extend out from the box in either direction to enclose the rest of the data or most of it. The whiskers stop when they come to the most extreme point that lies within 1.5 times the *inter-quartile range*, or IQR, of either end of the box.
- Outliers beyond the whiskers are indicated with individual markers.

Outliers are simply values that are distant from the bulk of the data. In drawing the boxplot, we defined them as values that were more than 1.5 IQR above the 75th percentile, or more than 1.5 IQR below the 25th percentile, but this was an arbitrary definition and you may see other definitions. More on outliers below.

Consider Figure 3.5, which is a boxplot of hospital sizes by number of beds in a hypothetical metropolitan area. We can glean the following information.

- Half the hospitals are between 99 and 351 beds.
- The IQR is 252 beds.
- The median hospital size is 189 beds.
- The rest of the hospitals are spread out between 50 and 604 beds, with the exception of four outlier hospitals.

Boxplots are a compact way to compare distributions. Figure 3.6 is a side-by-side boxplot comparison of the reduction in errors for the control and treatment hospitals.

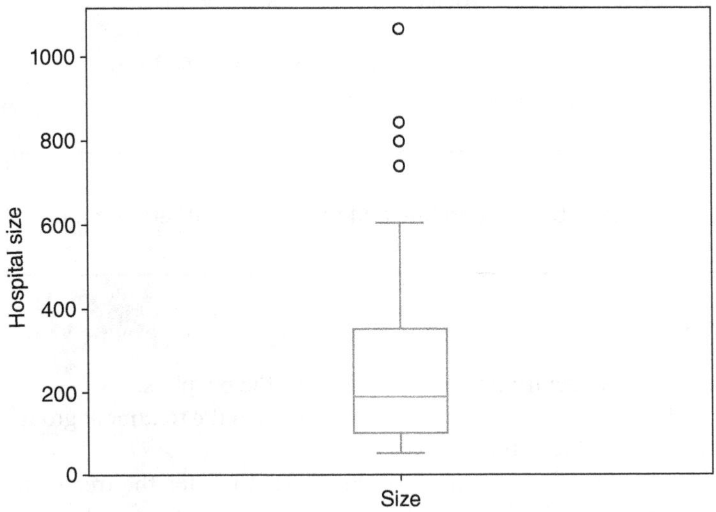

Figure 3.5 Boxplot of metropolitan area hospital sizes (y-axis shows number of beds).

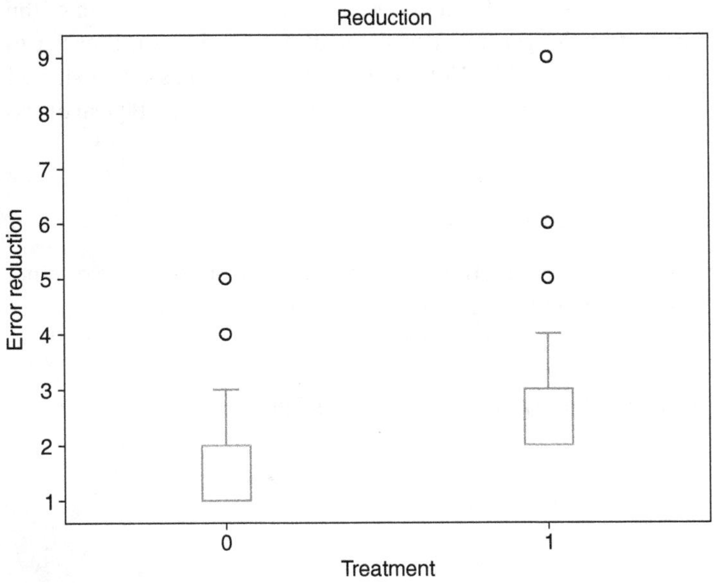

Figure 3.6 Error reductions (y-axis) for control hospitals(0) and treatment hospitals(1).

Different software has varying ways of displaying boxplots; some show the mean, others the median, some both.

Note how these boxes communicate information by the features that are missing—the median line and the lower whiskers.

> **Try It Yourself**
> What does the absence of the median line and the lower whiskers communicate?

3.6.5 Tails and Skew

Let's review the picture we get from the histograms and the boxplots.

The location of the data is lower for the control group than the treatment group, which is reflected in the value of the mean.

Other than the value of nine, the shape of the distribution for the treatment groups looks roughly like that for the control group. Both have peaks at the low end around one or two and trail off toward higher values. We call such a pattern *skewed toward high values*. The part of the picture where the data trail off, say around five, six, seven, eight, or nine, is called the *tail of the distribution*. The direction of the *skew* is the direction of the longer tail. The shape of the distribution is easier to see in the histogram than the table. Note how neither group has a bell-shaped ("Normal") distribution that is often (mistakenly) considered to be a typical distribution of data.

3.6.6 Errors and Outliers Are Not the Same Thing!

We saw how one hospital in the treatment group had an error reduction of nine, while all other hospitals, in both groups, were at six or below. We could consider this nine an outlier, and the data might merit a second look, but there is nothing inherent in this value that raises doubt.

However, suppose that Row 47 of Table 2.2 read as follows:

Row	Hospital#	Treat?	Reduction in Errors
47	4076	10	2

10 is not a valid value for the treatment/control variable, which needs to be either zero or one. We would then look up hospital number 4076 to see if we could find the reason for this error. (Perhaps it came during data entry, omitting a space between a 1 and a 0.)

Definition: Outliers A value (for a given variable) that seems distant from or does not fit in with the other values for that variable is called an *outlier*. It could be an illegal value, as in the above case. It could also be a very odd value or a legitimate one. If we saw a 456 in column four, this would not be illegal but it would be a very improbable degree of error reduction.

Some statistical software will identify outliers for you, but keep in mind that these are arbitrary identifications determined by arithmetic. Outliers are not necessarily errors—some are legitimate values. Consider these annual enrollments at a randomly selected set of 10 courses at Statistics.com.

8, 12, 21, 17, 6, 13, 29, 180, 11, 13

The 180 is certainly an outlier, but it is not incorrect. It is the enrollment in an introductory course, whereas the other enrollment figures are for more advanced courses.

Try It Yourself
Find the average enrollment for the 10 Statistics.com courses whose enrollments are listed above. Would you say this is a good representation of the typical enrollment?

Whenever we find an outlier, we need to investigate it and try to understand the reason for it. If there is an error, we need to try to correct it. Outliers, whether erroneous or legitimate, can strongly affect the numbers we compute from our data. In some cases, an outlier is a symptom of a deeper problem that could have an even greater impact on our results.

Outliers and Social Security
The US Social Security Administration (SSA) is a key source for wage data–Social Security taxes are due on almost all wages, and employers must file earnings reports with the tax authorities.

One statistic reported regularly is the average pay of those receiving more than $50 million in wages. This number receives a great deal of attention in the policy debate over income distribution. There were just 74 of these super-earners in 2009, and the government reported on October 15, 2010 that the average income of the super-earners more than quintupled in 2009–to an average of $519 million. This was quite an impressive feat during a severe recession, and the report came during a highly charged political atmosphere

(Continued)

(Continued)

in which an important bone of contention was the relative share of income and wealth held by the richest members of society.

Shortly after the report was issued, analysts found that two individuals were responsible for this entire increase. Between them, these two taxpayers reported more than $32 **billion** in income on multiple W2 (tax) filings. The SSA did not identify the individuals or reveal why they reported such astronomical sums.

However, the SSA did determine that the filings from the two individuals were in error and issued a revised report. The results?

- 2009 super-earner average wages actually declined 7.7% from 2008 instead of quintupling.
- 2009 average wages for all workers declined $598 from 2008; the original report was $384.

These two outliers had a huge and misleading impact on key government statistics. They contributed $214 to the average income of all wage earners, and when they were removed, the recession's wage hit grew by more than 50%. At the same time, the fuel they added to the income distribution debate was illusory.

What impact would these two outliers have had on statistics that used the median rather than the mean?

3.7 Python: Exploratory Data Analysis/Data Visualization

Data visualization is a core component of EDA. In this section, we will learn how to create data visualizations in Python using several packages. Given that most of these packages are built on top of the `matplotlib` library, it is useful to understand its key concepts to allow customizing the default data visualizations. All graphs in this book were created using Python and we often make use of customization. If the code is not shown in the book, you can find it in the Jupyter notebooks for each chapter available from the book's website at https://introductorystatistics.com/.

3.7.1 Matplotlib

`matplotlib` is a Python library for creating static, animated, and interactive visualizations in Python. It is a very powerful library that can be used to create a wide variety of visualizations. The `matplotlib` library is typically imported using the following command. `plt` is a common alias for `matplotlib.pyplot`.

```
import matplotlib.pyplot as plt
```

There are two different approaches to use `matplotlib` to create visualizations, the *pyplot* interface and the *object-oriented* interface. Here is an example of the two approaches to create a simple line plot. The resulting graphs are shown in Figure 3.7. Both approaches lead to the same result.

```
# create a data set
import numpy as np
x = np.linspace(0, 2*np.pi, 100)
y1 = np.sin(x)
y2 = np.cos(x)

# pyplot interface
plt.plot(x, y1) ①
plt.title("(a) pyplot interface")
plt.show()

# object-oriented interface
fig, ax = plt.subplots() ②
ax.plot(x, y1)
ax.set_title("(b) object-oriented interface")
plt.show()
```

① The `plt.plot` command creates a line plot by creating figures, x- and y-axes, and adding the line plot to the axes. Additional commands, like e.g. `plt.title`, operate on the current figure and axis.

② In the object-oriented interface, we use the `plt.subplots` command to explicitly create figure (`fig`) and axes (`ax`) objects. The `ax` object has several methods for manipulating the graph. For example, `ax.set_title`, sets the title for the axes.

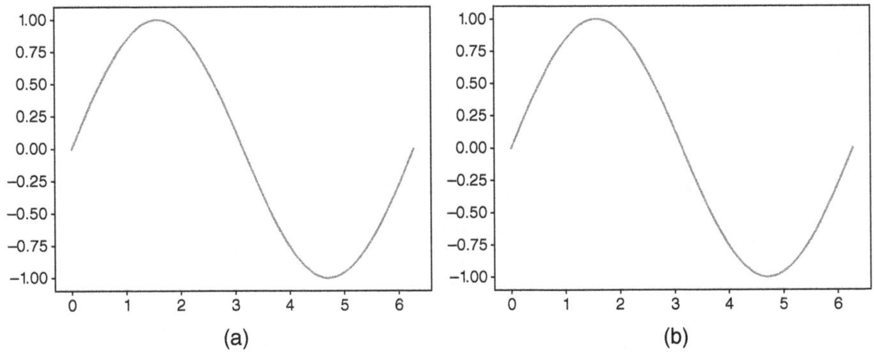

Figure 3.7 Simple line plot using the (a) pyplot and (b) object-oriented interface.

In this book, we prefer the object-oriented approach as it leads to cleaner, more explicit code. This is in particular true for visualizations with multiple graphs.

Useful commands for customizing the visualizations include, `set_xlabel`, `set_ylabel`, `set_title`, `grid`, and `legend`. The following code snippet shows how to use these commands to create a graph with a title, labels, and a legend. The resulting graph is shown in Figure 3.8. Note how `matplotlib` automatically chooses the colors for the two lines.

```
# create a figure
fig, ax = plt.subplots()
ax.plot(x, y1, color="C2", label="sin(x)") ①
ax.plot(x, y2, color="C3", label="cos(x)", linestyle=":") ②
ax.set_xlabel("x") ③
ax.set_ylabel("y = f(x)")
ax.set_title("sine and cosine") ④
ax.grid() ⑤
ax.legend() ⑥
plt.show()
```

① `label="sin(x)"` sets the text used in the legend

② `color="C3"` specifies the color of the line. `C0`, `C1`, …, refers to a pre-defined color palette that was chosen to be easy to distinguish. These colors are used by default. Alternatively, you can specify the color using a name, e.g. `color="red"` or using a RGB value like `color="#ff0000"`. For the

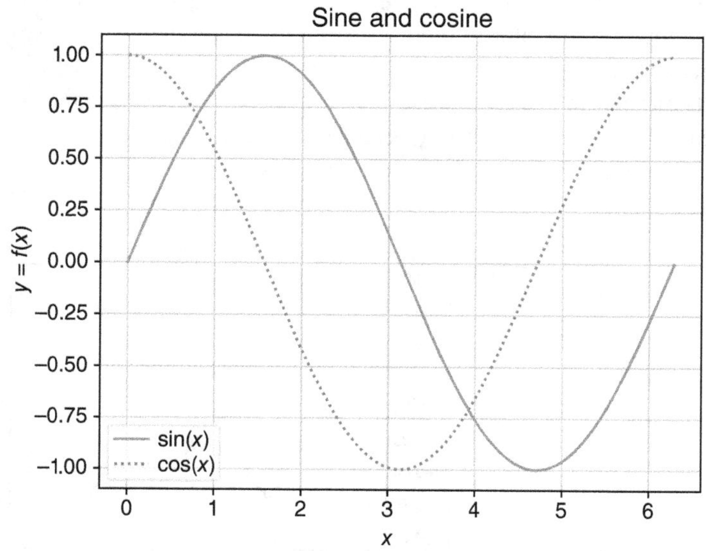

Figure 3.8 Customization of a graph with title, axes labels, legends and grid (`matplotlib`).

second line, we also specify the line style using `linestyle=":"` to display a dotted line.

③ `ax.set_xlabel("x label")` and `ax.set_ylabel("y label")` set the labels of the *x* and *y* axis.

④ `ax.set_title("title")` sets the title of the graph.

⑤ `ax.grid()` to add a grid to the graph. It can also be used to customize the grid further.

⑥ `ax.legend()` to add a legend to the graph and have finer control on what is included in the legend.

Two other useful methods are `ax.set_xlim` and `ax.set_ylim` to set the limits of the *x* and *y* axis. You can specify the minimum and maximum values as separate arguments, e.g. `ax.set_xlim(0, 1)`. An alternative is to specify the limits using a list, e.g.

```
limits = [ymin, ymax]
ax.set_ylim(limits)
```

3.7.2 Data Visualization Using `Pandas` and `Seaborn`

Many Python packages have their own visualization routines. For example, `pandas` has a *plot* method for dataframes and series. This *plot* method is a wrapper around `matplotlib`. The following code snippet shows how to use the `pandas` *DataFrame.plot* method to create a line plot of a dataframe and further customize it. The resulting graph will look similar to Figure 3.8.

```
# create a dataframe
import pandas as pd
df = pd.DataFrame({"x": x, "sin(x)": y1, "cos(x)": y2})
# create a line plot
ax = df.plot(x="x", y=["sin(x)", "cos(x)"]) ①
ax.set_title("sine and cosine") ②
ax.set_ylabel("y = f(x)")
plt.show()
```

① The `pandas` *plot* method creates a line plot of the columns `"sin(x)"` and `"cos(x)"` with the values of the column `"x"` on the *x* axis. The method returns an axis object `ax` which allows us to further customize the graph.

② Here, we add a title to the graph using the axis method *set_title*. The following line changes the axis label using *set_ylabel*.

Using the Python commands will create two separate lines that are only distinguished by color. If you need to support color-blind readers, you should use different line styles or markers. This can be done using separate plot commands for each line. The `pandas` plot functions also have an optional `ax` argument to add visualizations to an existing graph.

```
ax = df.plot(x="x", y="sin(x)")
df.plot(x="x", y="cos(x)", ax=ax, linestyle="-") ①
ax.set_ylabel("y = f(x)")
ax.set_title("sine and cosine")
ax.legend()
plt.show()
```

① We pass the axis object `ax` from the first `plot` command to the second DataFrame.plot command using `ax=ax`. This will add the second line to the same graph. We also specify the line style using the `linestyle` argument; `"--"` is shorthand for `"dashed"`

We can also use this to combine multiple graphs into one figure. Here is an example that creates two graphs next to each other (see Figure 3.9).

```
fig, axes = plt.subplots(ncols=2, figsize=(6, 3)) ①
df.plot(x="x", y="sin(x)", ax=axes[0]) ②
df.plot(x="x", y="cos(x)", ax=axes[1]) ③
axes[0].set_ylabel("sin(x)")
axes[1].set_ylabel("cos(x)")
plt.tight_layout() ④
plt.show()
```

① The `matplotlib` `plt.subplots` controls the size of the figure and the layout of axes. Here, we create a figure with a defined size (`figsize=(6,3)`) and with two axes next to each other. The `ncols` argument specifies the number of columns (2). Optionally, use `nrows` to define the number of rows. The plt.subplots function returns a tuple with the figure, the whole visualization, and the axes. The axes can be a single axis, a list of axes, or a list of lists of axes, depending on the number of rows and columns.

② By specifying `ax=axes[0]`, we add the plot of sin(x) to the left axis.

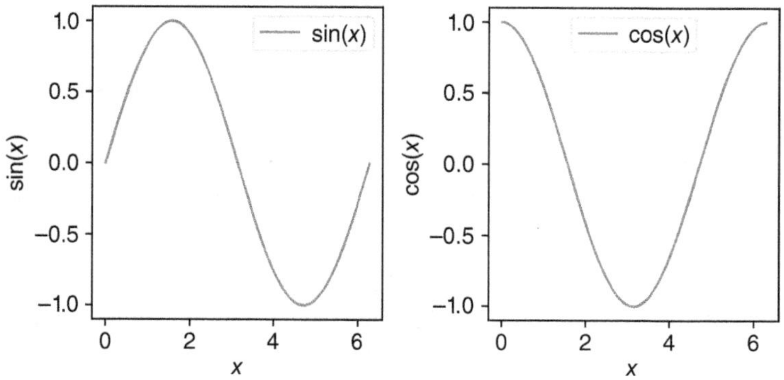

Figure 3.9 Creating two graphs next to each other.

③ This line adds the plot of cos(x) to the right axis.
④ The default settings for the layout often create axes that are too close together. The `matplotlib` function *plt.tight_layout()* separates the axes more.

Most figures in the book were created using this approach. Figure 3.10a shows the histogram from Section 3.6.2. The Python code for this graph is:

```
data = pd.read_csv("hospitalerrors_2.csv")
error_reduction = data[data["Treatment"] == 1]["Reduction"]
fig, ax = plt.subplots(figsize=(6, 3))
bins = [b + 1.5 for b in range(10)] ①
error_reduction.plot.hist(bins=bins, ax=ax, edgecolor="black") ②
ax.set_xlabel("Error reduction")
ax.set_ylabel("Number of hospitals")
plt.show()
```

① We explicitly set the bin edges (`bins`) to get bars that are centered around round numbers. Normally, we specify only the number of bins
② The *DataFrame.plot.hist* method creates a histogram for the values in the `error_reduction` series. The `bins` keyword argument usually takes an integer value that gives the number of bins. It is always a good idea to explore the effect of changing this number.

To compare two variables, we can use a scatterplot like the one shown in Figure 3.10b. The figure shows a relationship between payroll and team success in Baseball.

```
baseball = pd.read_csv("baseball_payroll.csv")
baseball.plot.scatter(x="Average Payroll (Million)", y="Total Wins") ①
ax.set_xlabel("Average Payroll (Million)")
ax.set_ylabel("Total Wins")
plt.show()
```

① The *DataFrame.plot.scatter* function creates a scatterplot graph.

Figure 3.10c shows the distribution of hospital sizes as a boxplot.

```
hospital_sizes = pd.read_csv("hospitalsizes.csv")
ax = hospital_sizes["size"].plot.box() ①
ax.set_ylabel('Hospital size')
plt.show()
```

① The *Series.plot.box* function creates the boxplot graph.

Figure 3.10d compares the distribution of error reduction with or without treatment using side by side boxplots.

```
data = pd.read_csv("hospitalerrors_2.csv")
fig, ax = plt.subplots(figsize=(5, 3.5))
axes = data[["Reduction", "Treatment"]].plot.box("Treatment", ax=ax) ①
axes["Reduction"].set_xlabel("Treatment") ②
axes["Reduction"].set_ylabel("Error reduction")
plt.show()
```

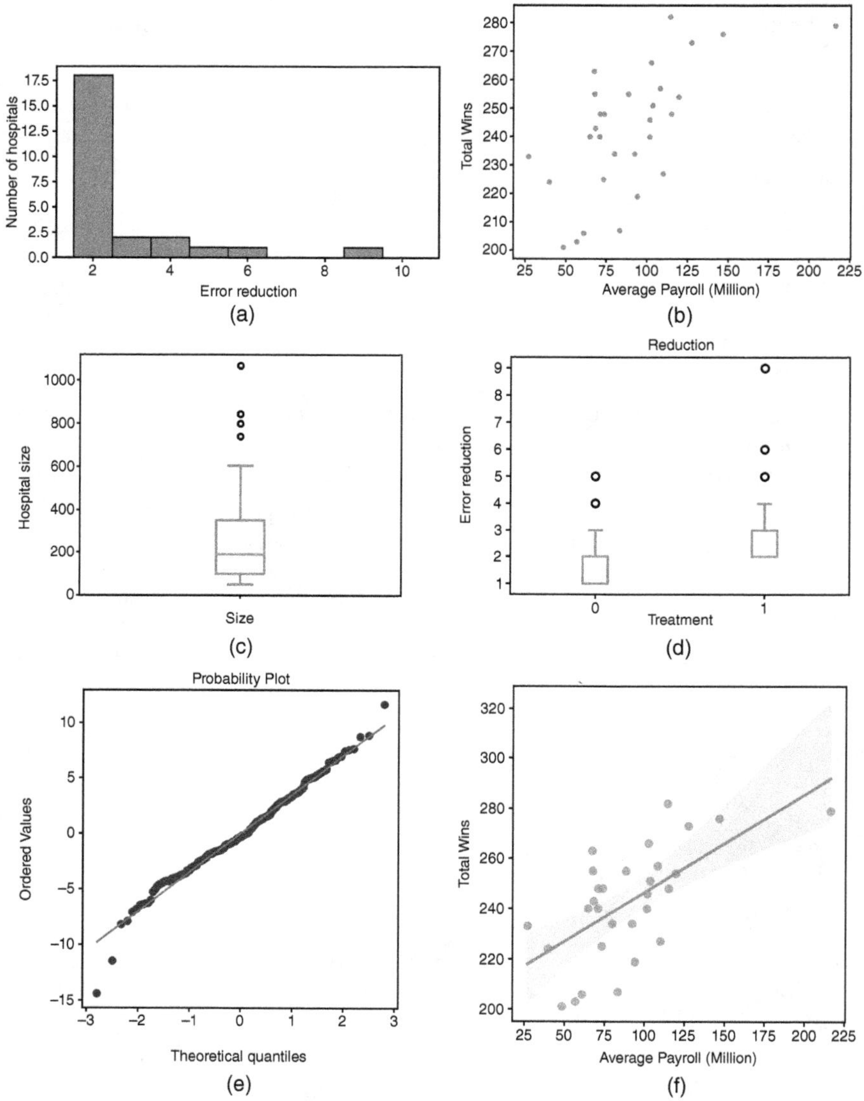

Figure 3.10 Examples of visualizations for exploratory data analysis created using Python. (a) Histogram with defined bins, (b) scatterplot to visualize how two variables are related, (c) boxplots give an overview of the distribution, (d) comparison of two distributions using boxplots, (e) QQ-plot for residuals, and (f) scatterplot with regression line.

① We first reduce the dataframe to the two columns that we want to visualize. The *Treatment* column is specified for grouping of the data in the boxplot. The function *DataFrame.plot.box* determines the unique *Treatment* values, here 0 and 1, splits the dataset into subsets and creates individual boxplots for each subset.
② The *DataFrame.plot.box* will create individual graphs for each of the remaining columns in the dataframe and returns the axes of the graphs. Here, we have only one column, so the list `axes` has only one value.

Figure 3.10e is a so called QQ-plot. It's a diagnostic plot to check if numbers follow a given distribution. The function *scipy.stats.probplot* is part of `scipy`.[2]

```
from scipy import stats
housing = pd.read_csv("boston-housing-model.csv")
fig, ax = plt.subplots(figsize=(5, 5)) ①
stats.probplot(housing["residual"], plot=ax) ②
plt.show()
```

① We define the figure size to create a square graph. This aspect ratio is preferred for QQ-plots.
② The *scipy.stats.probplot* creates the QQ-plot. By default, it compares to a normal distribution. Use the `dist` argument to specify a different distribution.

Another popular package for data visualization is `seaborn`. It is built on top of `matplotlib` and provides a high-level interface for creating statistical graphics. This means, we can create complex visualizations with just a few, often only one, statement. The following code snippet shows how to create a scatterplot with an overlaid regression line (see Chapter 10) using `seaborn`. This results in Figure 3.10f.

```
import seaborn as sns ①

baseball = pd.read_csv("baseball_payroll.csv")
fig, ax = plt.subplots(figsize=(5, 5))
sns.regplot(x="Average Payroll (Million)", y="Total Wins", data=baseball,
    ax=ax) ②
```

① The `import seaborn as sns` imports the `seaborn` package. The `as sns` part is a common alias for `seaborn`.
② The *sns.regplot* creates a scatterplot with a regression line. The `data` argument specifies the dataframe that contains the data and we use the column names to specify the *x* and *y* variables. Providing the optional `ax` argument allows us to add the graph to an existing figure and control the size and shape of the graph.

---

2 We will learn in Section 11.7.2 how to customize this plot further.

The *sns.regplot* not only plots the data as a scatterplot, but also fits a regression line to the data and estimates confidence intervals for it using bootstrap (light gray area); more about confidence intervals and bootstrap in Sections 7.3.1 and 7.5.

Exercises

3.1 Here are 20 more trials of the exercise you did at the beginning of the readings in which you investigated the model of a random distribution of 10 hospital errors between two years. Each row is a trial—10 coin flips. You will use the results to determine whether it is unusual for 10 hospital errors over two years to be split 7–3, just by chance.

Run		Run	
1	HHHTTHTTHH	11	TTHHHTTHHT
2	TTHHHTTTHH	12	TTTHHHHTHT
3	TTHHTHHTTT	13	TTHTHTTTTT
4	HTTHTHHTTT	14	THTHHHTTTT
5	HTHTHTHTTT	15	THTHHHTTTT
6	TTHTTTTHHH	16	HHTTHTHHHH
7	HHHTHTHHHH	17	HTTHTHTHTH
8	HHTHHHTTTH	18	THHTHHHHHT
9	HHTTTTHTTT	19	THHHTTHTTT
10	THHTTTHTTH	20	THTHHHTHTH

Each "H" or "T" represents an error. Under our chance model, let us say that "H" means the error happened in year 1 and "T" means year 2. Each row represents one trial (i.e. an allocation of 10 errors). Use Python to process the data.

a) For each of the 20 runs, count the number of times "H" (year 1) occurred.
b) Then make a frequency table for your results.
c) Visualize the results using a histogram.
d) What proportion of the runs yielded 7 or more "H's" (year 1)?
e) Comment on whether the difference between year 1 and year 2 might have happened by chance.

3.2 Suppose you are a consultant for a large garden supply retailer and you have been asked to determine whether to raise the price on a high quality hose product. How would you structure an experiment to answer this question…

a) In the online world

b) In the brick and mortar world

Be sure to incorporate the design principles outlined in this chapter, and/or discuss any problems involved in implementing them.

3.3 Estimate the probability that a family with 10 children would have three or fewer girls. Explain how you arrived at your estimate.

Hint: Assume that each successive child has a 50/50 probability of being male/female, and use dice, coins, or numbers drawn from hat.

3.4 Indicate whether the following statement is correct, along with a very brief explanatory statement: "In the Kerrich coin-tossing experiment, the proportions of heads and tails each tended towards 50% because, if an imbalance of in favor of heads developed, the probability of heads diminished slightly until balance was restored (and likewise for tails)."

3.5 A web consultant for a retail merchant tests two different presentations of a product to determine whether there is a difference in the length of time that people stay on that product page. She records the visit time in seconds per visitor for each presentation and finds that people viewing Presentation B (the new one) linger 0.33 seconds longer, on average, than those viewing Presentation A (the existing presentation). In the time allotted for the study, she is able to record 36 visits, 21 for Presentation A and 15 for Presentation B.

a) Load the data from dataset *web-page-data.csv* and calculate the average visit time for each presentation.

b) Create a barchart of the average visit time for each presentation.

c) Show the distribution of visit times for each presentation using a boxplot. What can you conclude from the visualization?

d) Create side-by side dotplots that show individual visit times for each presentation.

e) Histograms are a good way of showing the distributions of data. Create two visualizations that compare the distributions of visit times for each presentation.

i) Using a single graph, show both distributions using histograms of different colors.
   Hint: Use `alpha` to make the histograms transparent.
ii) Using two graphs arranged in rows, show the distributions in separate histograms. Make sure that the axis range for the visit times is the same in both graphs.

Compare the two visualizations. Which one do you prefer? Discuss the advantages and disadvantages of each visualization.

4

Accounting for Chance–Statistical Inference

In this chapter we connect probability to the process of drawing conclusions from data. After completing this chapter, you will be able to:

1) Explain the rationale for why hypothesis tests are needed
2) Identify when hypothesis tests are appropriate
3) Distinguish the circumstances that call for formal vs. informal hypothesis tests
4) Explain the logic of a hypothesis test
5) Interpret the results of a hypothesis test

The task of trying to assess the impact of random variability on the conclusion from a study, or the results of a measurement, is called *statistical inference*. In this chapter we will look at a particular kind of statistical inference called a *hypothesis test*. Generally, a hypothesis test seeks to determine whether the effects we see in some data from a study are real or might just be the result of chance variation. The logic of a hypothesis test runs as follows:

Variation from random chance is everywhere, and we are easily fooled into thinking it might be meaningful. When we see a data pattern or effect that we think is important and real, e.g. a new medical treatment produces better health outcomes or a new web ad being tested produces more clicks, we set up a chance model (flipping coins or drawing numbers) to see whether it can produce a result as unusual as the pattern or effect that we actually saw.

4.1 Avoid Being Fooled by Chance

Why does hypothesis testing, perhaps the most confusing and controversial aspect of statistics, exist? Hypothesis testing is a way of protecting yourself against being fooled by chance.

There is a powerful need for the human brain to explain the world. Data that comes in to the eyes or ears is fitted into seemingly meaningful patterns.

Statistics for Data Science and Analytics, First Edition. Peter C. Bruce, Peter Gedeck, and Janet Dobbins.

Companion website: www.wiley.com/go/Wiley_Statistics_for_Data

This happens even when the data are produced by random chance—we tend to discount the ability of chance to generate patterns that look meaningful and real. Diseases are often complex and have varied outcomes if left on their own, yet we tend to attribute health changes to recent changes in diet, activities, or environment. Sports commentators import undue meaning to short-term periods of good or bad performance. A couple of big sales will send a marketing manager scurrying to find out what produced them (and a corresponding small dip in sales will send him in search of the cause).

Try It Yourself
In the book resources website https://introductorystatistics.com/, take a look at the "50 Coin Tosses" experiment. One group of people was told to toss 50 coins and record the results. Another group was told to make up 50 imaginary coin tosses and write down the results. Can you tell which group is which?

Who uses hypothesis testing?

- The *research* community uses hypothesis testing to determine whether a study is worthy of regulatory approval or publication (its role as gatekeeper in the latter is increasingly controversial).
- *Data scientists* use it to assess the results of their tests and models. They have less need of the formal apparatus of hypothesis testing, but do use the resampling methods presented here, and their variants, to help separate random from real patterns in data.

4.2 The Null Hypothesis

The standard hypothesis-testing procedure involves a what-if calculation. We ask, "Could my study results or the results of my analysis be explainable by chance?" This supposition is called the *null hypothesis*. Here, "null" means nothing important is really happening, and whatever differences or variations we observe between groups are just due to chance. Then we calculate how big an effect chance might have in our situation, under the assumption that nothing unusual is going on. If the real-world result that we observe is consistent with the range of outcomes that chance might produce, then we say that what we observed may well be due to chance. In other words, if our observations could be due to chance, we do not reject the null hypothesis. However, if what we observe is much more extreme than what we would expect due to chance, then it is likely that something else is going on.

4.3 Repeating the Experiment

One way to check the results of our experiment would be to repeat it over and over to see if the result holds. However, doing just one experiment takes a lot of time and money. Repeating it multiple times is out of the question. So instead we repeat it on a computer, using the model suggested by the null hypothesis, the *null model*. After all, if the null model is valid, it should be able to produce results like the ones we actually observed.

If the null model is valid and our results were just due to chance, that would mean that the apparent superiority of the hospital treatment group in Chapter 2 was just due to the luck of the draw when we assigned hospitals to treatments. If we assigned them differently, the results would change. Well, let's assign them differently to see how much they might change. The null model suggests we should assign them randomly.

4.3.1 Shuffling and Picking Numbers from a Hat or Box

A good way to visualize the process of randomizing the hospital errors and reassigning them is to imagine picking numbers from a hat or box.

1) Write the error reduction scores from the 50 hospitals on slips of paper and put them in a hat or box.
2) Shuffle the papers in the hat and deal them out into two "resampled" groups of 25.[1]
3) Calculate the mean error reduction in each of the two groups, and record the difference between them.
4) Repeat steps 2 and 3 many times.
5) Find out how many resampled differences exceeded the observed difference.

> **Try It Yourself**
> Stop now and reflect on just one shuffling of the hat. What is your best guess about the difference in mean error reduction between the two groups of 25 drawn from the hat?

Definition: Permutation Test Combining two or more samples in a hat or box, shuffling the hat, then picking out *resamples* at random is called a *permutation test* (sometimes called a "randomization test"). If there were two original samples

1 Sometimes each paper is replaced and the hat reshuffled before *each* draw–resampling *with replacement*. With replacement and without replacement have slightly different statistical properties, but produce results that are sufficiently similar for our purposes.

combined in the hat, then you typically draw out two resamples. If there were n original samples in the hat, then you would draw out n resamples.

Hat Video
The accompanying website for this book contains a video of a permutation test using the hospital error reduction data. Watch the video to see a couple of trials of this permutation test, then continue with the computer implementation below.

Table 4.1 is one reshuffling of the medical error reduction scores done by a computer.

The means of the two groups differ, but not by as much as before. Now they only differ by 0.36. That difference is one estimate of how big a difference might be due just to chance. It would be helpful to have another estimate. In fact, since the computer is doing all the work, why not 50 more? Table 4.2 shows the values of the differences in average error reductions for 50 different permutations.

This is encouraging. We never get a difference as large as 0.92 between the two groups as a result of reshuffling.

Try It Yourself
What do these results mean?

***P*-value**
Previously, we defined the p-value as the probability that the chance model might produce an outcome as extreme as the observed value. In this simulation, the chance model never produced a difference as large as 0.92, so the estimate of the p-value is 0. But we only did 50 trials. If we did 10,000 trials we might well see a chance result as large as 0.92, which would raise the estimated p-value slightly above 0.

4.3.2 How Many Reshuffles?

How many shuffles do we need to get good accuracy for the p-value? We just tried 50 and found that the observed result of 0.92 seemed pretty unlikely. Let's now try 1000. Table 4.3 shows the sorted results from the first 15 rows of 1000 trials.

Only 9 trials out of 1000 (bold type in Table 4.3) had differences in their means of 0.92 or more. Figure 4.1 shows the histogram of all 1000 trials.

Table 4.1 Permutation of error reduction scores into two groups.

	Random Group 1	Random Group 2
	1	1
	4	2
	1	1
	2	3
	2	1
	3	2
	1	1
	1	2
	2	1
	2	1
	2	4
	5	1
	2	1
	2	6
	2	3
	2	2
	2	9
	4	2
	2	2
	2	3
	2	2
	2	2
	2	5
	2	2
	2	4
Mean	2.16	2.52

4.3.3 The *t*-Test

Repeated shuffling is no problem for today's computers, but it was prohibitively costly before the computer age. Mathematical approximates to the resampling distribution were worked out and used instead. The mathematical approximation that is used in assessing an A/B test with numerical data is the "*t*-test," based on the distribution of the *t*-statistic. We will learn more about the *t*-distribution and how it was developed in Chapter 7.

Table 4.2 Average error reduction: first random group–second random group, 50 trials (sorted by difference).

−0.76
−0.68
−0.68
−0.60
+ 45 more
0.60

Table 4.3 Sorted results from 1000 trials (first 15 shown).

1.56
1.44
1.08
1.08
1.00
0.96
0.92
0.92
0.92
0.88
0.88
0.88
0.88
0.84
0.84
+ 985 more

4.3.4 Conclusion

Our chance model produced the observed result only rarely, and this was consistent across multiple simulations of 1000 trials each. If you were to count up the extreme (>=0.92) results from all the simulations, they would account for less than

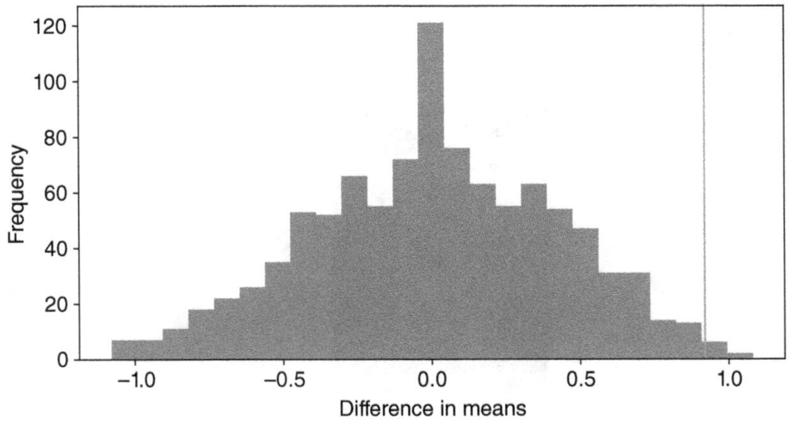

Figure 4.1 Histogram of 1000 trials—permutation of control and treatment groups (x-axis is the mean of the first shuffled group minus the mean of the second shuffled group).

2% of the total. This is sufficiently rare to reject the chance model. Two questions, though, may remain in your mind.

1) How rare does rare have to be to reject the chance model? In other words, how low should the p-value be?
2) Is there a firm rule we can use to determine the number of simulations required?

The answer to question one is somewhat arbitrary. If you set the bar for the p-value at a very low value, such as 1%, you'll have good protection against being fooled by chance. However, you'll also miss more real events because they will appear to be within the range of chance variation. If you set the bar relatively high, such as 10%, more events will qualify as real, but there is a greater probability that some of them are just the result of chance. There is no magic right answer, though standard practice is to use a threshold of 5%.

Origin of the Five Percent Rule

R. A. Fisher, one of the "founding fathers" of statistics, is generally regarded as the source of the prevalent standard that a p-value of 5% is the threshold of "statistical significance." He thought that if the chance model can produce results as extreme as the observed data 5% of the time or more, we can't rule out chance. Source: The University of Adelaide/Wikimedia Commons/Public domain.

(Continued)

(Continued)

Alcoholic beverages are legal in the United States, but it is not legal to drive while under the influence of alcohol. In most areas, there are legal limits that define "under the influence" in terms of a certain cut-off for blood alcohol level. What are the pros and cons of setting this cut-off at a high level? At a low level?

Question two also has no right answer. The number of simulations required depends on the purposes of the investigation.

- If you are a researcher investigating a new drug, the clinical trial will cost hundreds of millions of dollars and the drug regulatory agency will require a rigorous presentation of the results. If you are publishing a paper in a scientific research journal, you will be asked to justify how you arrived at your conclusion. In either case, you might use a resampling procedure (perhaps with thousands or hundreds of thousands of simulations), or you might use a formula calculation to show what the resampling simulation would resolve to "in the limit."
- If you are a data scientist examining some data to assess the possible role of chance, perhaps as a preliminary step before further action, a high level of precision is not required.

As a final step in our example, we did 20,000 trials. Of those, 248 had differences in means as great as 0.92 for a *p*-value of 248/20,000 = 0.0124. This confirms the earlier trial results and the conclusion that the chance model is not likely to produce the observed value.

Definition: Statistically Significant The results of an experiment or other study are statistically significant if they are too unlikely to be accounted for by our chance model.

4.4 Statistical Significance

How unlikely does our outcome have to be before we reject the null hypothesis and declare our result statistically significant? This is partly for the analyst to decide, and there is no definitively right or wrong answer. Whatever we choose for the answer, the mathematical symbol for it is the Greek letter alpha, α.

Definition: α (alpha) Alpha is the decision threshold you set for the *p*-value in advance of an experiment. For example, you may decide that a *p*-value less than 0.05, a common choice, lets you rule out chance.

What is a reasonable value for α depends on the situation. If we are really picky and do not reject the null hypothesis unless α is less than 0.00000001, we will rarely claim a real difference by mistake when there is none—this is called a Type I error. On the other hand, we will frequently miss a real effect by not rejecting the null hypothesis. This is called a Type II error. In practice, setting a reasonable alpha depends on the relative costs for the two types of errors.

Definition: Type I Error When you erroneously conclude that an effect is real when it is just chance, you have committed a Type I error. This occurs when you get a very low *p*-value, which indicates a low probability of the result happening by chance, but the result is, nonetheless, still due to chance.

Definition: Type II Error When you conclude that an effect could be due to chance although it is real, you have committed a Type II error. This occurs when the effect is real, but due to chance and small sample size, you get a *p*-value that is not low enough.

4.4.1 Bottom Line

When you do a study and make decisions about whether results are interesting and statistically significant, you are balancing the risks and costs associated with these two types of error. For example:

- For the hospital error illustration, the social cost of a Type I error would be to implement a program that really does no good. The main financial cost would be the program cost itself. The social cost of a Type II error would be that we discard a helpful program. Then the cost is measured less in dollars and more in the unnecessary harm due to errors.
- For a web dating service testing, whether a slimmed-down registration form yields more registrants, the cost of a Type I error (and switching to the new form) would be a loss of subscriber data, which is important in creating matches. The cost of a Type II (sticking with the old form) error would be loss of the additional subscribers that would be generated by the new form.

Getting back to the hospitals, if this is just the first stage of research on a program, we would probably be inclined to set alpha moderately high so that if there is even a mild sign it is helpful, we can do further research. Perhaps an alpha of 10% would be sensible. An alpha of 50% would almost always be too high. That would mean accepting as unusual and acting upon things that happen randomly rather than due to the treatment or other change that we make.

Some researchers advocate setting a level for alpha in advance. The idea is that you set the rules before playing the game. Then you can't decide at the end of the game to let alpha be whatever it needs to be for you to win.

4.4.1.1 Statistical Significance as a Screening Device

The idea of statistical significance is sometimes used outside the framework of formal hypothesis testing as a screening device. For example:

- Some machine learning algorithms that perform many (sometimes millions) of operations on data in search of patterns use statistical significance tests as guidance, for example what variables to include in a model. These decisions are made repeatedly, on the fly, in automated fashion, and are somewhat arbitrary in the sense that there is no attempt made to interpret individual hypothesis tests.
- An agricultural research station may test numerous varieties of corn in one summer and use *p*-values as a threshold for proceeding with further development of a limited set. The costs of both Type I and Type II error are not huge, and there are no regulators or journal editors peering over your shoulder, so the application of *p*-values as a threshold can be relaxed and informal.

Another type of screening takes place when scientists submit research for publication. In scientific journals, it is very hard to get published with an alpha bigger than 5%, so most authors just use that. Indeed, many researchers probably do not fully understand what statistical significance is all about, and simply view it as a hurdle to jump over in getting published.

***P*-Hacking**

The use of *p*-values became controversial in 2015, when the so-called practice of "*p*-hacking" gained notoriety. *P*-hacking is the process of submitting data to numerous statistical tests in the hopes that one test, benefiting from the natural random variability of data, will prove to be "statistically significant." *P*-hacking can include the following:

1) Trying different test statistics (metrics)
2) Testing different subgroups of the data
3) Looking backwards to select different time points to capture the data

4) Using different sources for similar data

Statistically significant p-values are the gateway to publication for academics, so a motivation for misuse comes naturally. One psychology journal "banned" the use of p-values, and the American Statistical Association felt compelled to issue a policy statement on the use of p-values.

4.4.2 Torturing the Data

A phenomenon related to p-hacking is the practice of "data dredging:" searching through a large store of data, or through different data sources, in search of interesting patterns (patterns that are often, in reality, just the result of chance). The economist Ronald Coase had this in mind when he said, "If you torture the data long enough, it will confess to anything." Tyler Vigen's book, *Spurious Correlations*, illustrates a number of such examples. For example, he shows in a plot how suicides by hanging move in tandem with US spending on scientific research (see tylervigen.com). Sometimes these spurious effects are the unintended byproduct of automated machine learning procedures that churn through large amounts of data in search of patterns, but fail to adequately validate results with fresh data. John Elder, founder of a data science consultancy, terms this the "vast search effect." Other times these patterns emerge when an analyst hopes for a certain result and keeps looking until it is found.

The Vast Search For Mules

The movie *2000 Mules*, concerning allegations of fraud in the 2020 US presidential election, illustrates the vast search effect.

(Continued)

(Continued)

The movie's producers combed through 10 trillion cell phone pings to locate 250 cell phones it claimed belonged to people ("mules," as in carriers for drug dealers) supposedly delivering fraudulent ballots. The supposed evidence? Their phones were frequently in the vicinity of ballot drop boxes. The *Wall Street Journal* (July 23, 2023) noted, "With 10 trillion pings, what are the odds that some random patterns would show up?" The answer is "pretty high:" with 10 trillion pings, you could find almost any location pattern you want. In 2024, the movie's producers disavowed the film and apologized. Source: Wery Shania/Pexels.

4.4.3 Practical Significance

Statistical significance must be contrasted with practical significance—whether the difference is big enough to make a practical difference outside the study in the realm of business, medicine, etc.

- The treatment in our experiment reduced errors by almost 50% (about one per hospital). These results definitely seem to have practical significance.
- Now consider a web marketing experiment for a digital streaming service with \$100 million in revenue and \$95 million in costs. A new email reminder (different from the existing reminder) produced 3% more subscription renewals. 3% sounds small, but revenue is now \$103 million as a result. Additional subscriptions can be handled at little cost; costs rise only \$1/2 million. With the 3% boost in revenue, profit has gone from \$5 million to \$7.5 million, a 50% increase—of definite practical significance. However, the 3% change would need to be based on a much larger sample than in our hospital study to be distinguishable from chance—i.e. *statistically significant*.

With a large enough sample, any difference, no matter how minor and unimportant, can become statistically significant. Generally, the ability to detect differences increases with the number of units studied. Statistical power is a measure of our ability to detect a real effect.

When reporting the results of a hypothesis test, we should report both whether the result was statistically significant and just how big it actually was—the effect size. So for the errors experiment, we would report that we found a reduction of 0.92 in the number of errors as a result of the treatment and that we believe that difference was not due to chance. If possible, a report should include an evaluation of the practical importance of the size of the effect that we observed. Such an evaluation should be done by someone familiar with the field under study—i.e. a medical administrator for the hospital study, and a marketing manager for the digital subscription experiment.

4.5 Power

If the result of your hypothesis test is "not statistically significant," can you conclude that the effect you are looking for does not exist? No, you may simply have had too small a sample to detect it—the sampling variability was too broad. The ability of a given sample size and hypothesis test to detect an effect, if one exists, is termed *power*.

Definition: Power Power is the probability that a statistical test will identify a specified "effect," i.e. determine that there is a statistically significant difference when one exists. For example, you might want to determine the sample size needed to detect a 5% improvement in a product's "Net Promoter Score" (derived from surveys asking whether a consumer would recommend the product to others). To calculate power, you need to know (1) the effect size you want to discern, (2) the sample sizes, and (3) something about sample variances and distribution.

Calculations for power can be complex and are typically handled by software algorithms developed for this purpose. Calculating power is important, however, before undertaking a costly study. You want to be sure that the sample you are planning to collect is big enough to detect the effect that you are interested in identifying.

4.6 The Normal Distribution

Let's use our results to briefly introduce a concept that has been central to statistics for a century. The histogram of our first 1000 reshuffles, shown in Figure 4.1, looks a lot like a theoretical distribution called the Normal distribution. The Normal distribution is bell-shaped and symmetric—it has identical tails on both sides. Figure 4.2 shows what the theoretical Normal distribution looks like.

The *x*-axis represents the metric of interest, the *y*-axis measures the relative frequency of the scores (more about this later).

The Normal distribution was originally called the error distribution—it depicted not the distribution of the original data, but of *errors* or *deviations* from predicted values. The term "Normal" came into use over time to describe typical distributions of these errors. In fact, the original raw data encountered in research and business are typically *not* Normally distributed, but measurement and estimation errors, as well as statistics calculated from samples, are more likely to be Normally-distributed. We will learn more about the Normal distribution in the next chapter.

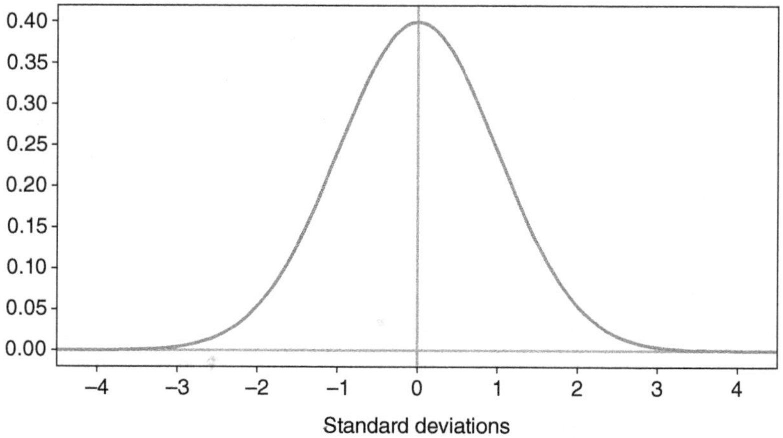

Figure 4.2 The theoretical Normal distribution (the x-axis is expressed in standard deviations).

> **Most Data Are Not Normal**
> Most naturally-occurring (raw) data are not Normally-distributed. For example:
>
> - Heights: On average, men are taller than women, children are shorter than adults, and different ethnic groups can vary in their heights. Only when you control such factors and restrict the definition of the group, e.g. black American adult females, does the distribution turn Normal.
> - Incomes: The distribution of incomes is long-tailed or skewed to the right to account for the individuals who have incomes many times greater than the average.
> - Tech Support Call Duration: Like incomes, this distribution is right-skewed, reflecting the fact that there are a few extremely difficult and lengthy calls.

4.6.1 The Exact Test

Before we leave this subject, let's look at one more approach. Instead of simply shuffling the error reduction values, we could examine all 126, 410, 606, 437, 752 ways of arranging the 50 numbers into two groups of 25. These 126, 410, 606, 437, 752 numbers are the complete permutation distribution. Then we would compare the observed result to this distribution. Obviously it is going to take some computing power to go that route, but with modern machines and clever algorithms, this is a viable approach in some situations. Hypothesis tests done in this way are known as *exact tests*. Instead of exact tests, it is a lot easier to just try 1000 or 5000 of the possible arrangements, which is what we did. Trying such a random subset of the possible arrangements instead of all of them is how permutation tests are usually carried out.

4.7 Summary

The human mind can have difficulty with chance events, interpreting them as meaningful. One way to protect against being fooled by chance patterns in data is the hypothesis test, in which we postulate a chance model and determine how often that chance model can produce a result (for example, a difference between treatment and control) as extreme as the observed result. The hypothesis testing process can be very formal, as with regulatory submissions and academic publications. Or it can be less formal, to provide perspective on statistical or machine learning analysis.

4.8 Python: Random Numbers

The `random` package, which is part of the Python standard library, provides functions to generate random numbers.[2] While it is called random, the numbers are pseudo-random, meaning that they are generated by a deterministic algorithm. This means that random numbers can repeat after a certain period. With this drawback in mind, pseudo-random numbers are sufficient for most applications. Even more, they also allow for reproducibility by creating the same sequence of random numbers. This is important for statistical research.

4.8.1 Generating Random Numbers Using the `random` Package

The `random` package uses the Mersenne twister by default. This random number generator (RNG) can generate $2^{19937} - 1$ different numbers before it repeats itself. To create a random number, use the *random()* function. This function returns a random float number between 0 and 1.

```
import random
print(random.random(), random.random())
print(random.random(), random.random())
```

Output

```
0.8515514062887293 0.9099953287797414
```

2 The `random` package is sufficient for modeling and simulation. For security or cryptography, use packages like `secrets` that create cryptographically strong random numbers.

```
0.8681448728121443 0.1420894518957383
```

The result of this call will change every time you run it. To get the same result, we can seed the RNG with a fixed value. This is useful for reproducibility.

```
random.seed(123)
print(random.random(), random.random())
random.seed(123)
print(random.random(), random.random())
```

Output

```
0.052363598850944326 0.08718667752263232
0.052363598850944326 0.08718667752263232
```

In addition to this simple function, the `random` package provides a number of other functions to generate random numbers. Here is an overview of the most important functions.

- *randrange(start, stop, step)*: Returns a random integer between `start` and `stop` (exclusive) with a given `step` size.
- *randint(a, b)*: Returns a random integer between `a` and `b` (inclusive).
- *choice(x)*: Returns a random element from the given list `x`.

For example:

```
print(random.randrange(0, 10, 2)) # 6 - output from 2, 4, 6, or 8
print(random.randint(0, 10))      # 4 - output from 0, 1, ..., 9, or 10
print(random.choice([0, 1, 2]))   # 0 - output from 0, 1, or 2
```

There are also three function that operate on lists:

- *shuffle(population)*: Shuffles the list `population` in place.
- *sample(population, k)*: Returns a list of `k` elements from the given `population` **without** replacement.
- *choices(population, k=1)*: Returns a list of `k` elements from the given `population` **with** replacement.[3]

For example:

```
x = [0, 1, 2, 3, 4, 5, 6, 7, 8, 9]
random.shuffle(x)
print(x) # [7, 8, 3, 4, 1, 9, 2, 5, 6, 0]
print(random.sample(x, k=3)) # [4, 3, 7]
print(random.choices(x, k=6)) # [0, 5, 6, 4, 7, 5]
```

3 The full signature of the *choices* function is `choices(population, weights=None, *, cum_weights=None, k=1)`. The additional `weights` and `cum_weights` arguments allow to specify a probability distribution of the elements of the population. Note the use of the `*` in the signature, which indicates that the arguments after it are keyword-only arguments. You always need to specify the name of the argument when calling the function. It is good practice to use the keyword name in calls irrespective of it being enforced or not.

You will find more examples for these three functions in many places in this book.

The random package also has functions to generate random numbers from a number of probability distributions. For now, here are three examples. We will learn more in Section 5.6.

- *random.uniform(a, b)*: Returns a random number from a uniform distribution between a and b.
- *random.gauss(mu, sigma)*: Returns a random number from a Normal distribution (also called Gaussian) with mean mu and standard deviation sigma.
- *random.expovariate(lambd)*: Returns a random number from an exponential distribution with rate lambd.

4.8.2 Random Numbers in numpy and scipy

You can also generate random numbers with the numpy package. It implements several different random number generators. The default is the Permuted Congruential Generator (PCG64). In newer releases of numpy, it is recommended to use the numpy.random.Generator class instead of the functions in the numpy.random module. It is faster and can be used in parallel applications.[4]

```
import numpy as np
rng = np.random.default_rng(seed=321)
print(rng.random()) # 0.6587666953866232
print(rng.random()) # 0.9083083579615744
```

Like the random package, numpy has functions to sample random integers or real numbers from a defined range.

```
print(rng.integers(low=0, high=10, size=3)) # [6 4 3]
print(rng.uniform(low=0, high=10, size=3))
# [5.55856507 9.00575242 6.82980572]
```

The function *rng.choice* allows to sample from a list of elements with or without replacement. The default is sampling with replacement.

```
# sample with replacement
print(rng.choice([0, 1, 2, 3, 4], size=3)) # [3 3 0]
# sample without replacement
print(rng.choice([0, 1, 2, 3, 4], size=3, replace=False)) # [2 1 0]
```

The numpy.random module also provides functions to sample from a number of probability distributions. Here are two examples:

4 We will use the new approach throughout this book. If you use a package that depends on the legacy version of the RNG, you can set a random seed using np.random.seed(123).

- *rng.normal(loc=0.0, scale=1.0, size=None)*: Returns a random number from a Normal distribution with mean `loc` and standard deviation `scale`.
- *rng.exponential(scale=1.0, size=None)*: Returns a random number from an exponential distribution with scale `scale`.

If you set the `size` argument, you can sample multiple numbers at once. The result is a `numpy` array.

The `scipy` package gives access to even more distributions. In addition to sampling from these distributions, it also provides functions to calculate the probability density function (PDF) and the cumulative distribution function (CDF). Let's look at the Normal distribution as an example:

```
from scipy.stats import norm
# create a RNG with a fixed seed for reproducibility
rng = np.random.default_rng(seed=123)
print(norm.rvs(loc=0.0, scale=1.0, size=3, random_state=rng))
# [-0.98912135 -0.36778665  1.28792526]
print(norm.pdf(x=0.0, loc=0.0, scale=1.0))
# 0.3989422804014327
print(norm.cdf(x=0.0, loc=0.0, scale=1.0))
# 0.5
```

4.8.3 Using Random Numbers in Other Packages

Many other statistical or data science packages make use of random numbers. To make your results reproducible, it is useful to understand which method is used and how to set a random seed in these packages. For example, the `pandas` package, like many other packages, uses `numpy` to generate random numbers. Here is Python code to sample from a `pandas DataFrame`:

```
import pandas as pd
df = pd.DataFrame({"a": [1, 2, 3, 4, 5], "b": [6, 7, 8, 9, 10]})
print(df.sample(n=3, random_state=123))
```

Output

```
   a   b
1  2   7
3  4   9
4  5  10
```

If we don't set the `random_state` argument, the result will change every time. The `pandas` package uses the `numpy` RNG, so you could also use the generator `rng` from the previous example. Alternatively, you can set the random seed using, e.g. `np.random.seed(123)`.

4.8.4 Example: Implement a Resampling Experiment

Let's look at an example. In this chapter, we discussed the reduction of errors in hospitals. We first need to load the data from the file `hospitalerrors_2.csv` using `pandas`.

```
import pandas as pd
df = pd.read_csv("hospitalerrors_2.csv")
print(df.head())
```

Output

```
   Row  Hospital  Treatment  Reduction
0    1       239          0          3
1    2      1126          0          1
2    3      1161          0          2
3    4      1293          1          2
4    5      1462          1          2
```

We are interested in the reduction in errors for the treatment group.

```
mean_reduction=df[["Treatment", "Reduction"]].groupby("Treatment").mean() ①
print(mean_reduction)
```

① The *DataFrame.groupby()* method groups the data by group (treatment or control). In a subsequent step, `mean` calculates the mean of the *Reduction* values of *each* group.

Output

```
           Reduction
Treatment
0               1.88
1               2.80
```

The observed difference is

```
observed_difference = (mean_reduction.loc[1, "Reduction"] -
                       mean_reduction.loc[0, "Reduction"]) ①
print(f"Observed reduction {observed_difference:.3f}")
```

① Use *DataFrame.loc* to access individual cells of a dataframe. The *loc* method uses the index values to identify the cell.

Output

```
Observed reduction 0.920
```

We can now implement the resampling experiment. There are many ways to do this. Let's first have a look at using pure Python. We extract the observations (`Reduction`) and the treatment (`Treatment`) information.

```
observation = df["Reduction"]
treatment = df["Treatment"]
```

Using the random package, we create a shuffled version of the observations.

```
import random
random.seed(123) ①
shuffled = list(observation) ②
random.shuffle(shuffled)

# split the shuffled observations by treatment group
observed_0 = []
observed_1 = []
for obs, treat in zip(shuffled, treatment): ③
    if treat == 0:
        observed_0.append(obs)
    else:
        observed_1.append(obs)

# calculate the mean reduction for the treatment and control group
obs_treatment_0 = sum(observed_0) / len(observed_0) ④
obs_treatment_1 = sum(observed_1) / len(observed_1)

# calculate the difference
obs_difference = obs_treatment_1 - obs_treatment_0
print(f"Observed difference after shuffling: {obs_difference:.3f}")
```

① Set the random seed for reproducibility.
② This creates a copy of the list to make sure we don't modify the original list with the next *random.shuffle*.
③ The *zip* function allows to iterate over two lists at the same time. It returns a sequence of tuples, where each tuple contains the elements of the two lists at the same position.
④ The Python function *sum* adds up all values of a list. Dividing by the length of the list (the number of elements of the list), calculates the mean of the list.

Output

```
Observed difference after shuffling: 0.200
```

There are many ways to write this code more concisely. For example, we could use list comprehensions to create the two lists observed_0 and observed_1.

```
observed_0 = [obs for obs, treat in zip(shuffled, treatment) if treat == 0] ①
observed_1 = [obs for obs, treat in zip(shuffled, treatment) if treat == 1]
```

① List comprehensions are a great way to convert loops into a single, concise statement. More about list comprehensions in Section 2.18.6

Or we make use of the numpy function *np.mean* to calculate the means:

```
import numpy as np
obs_treatment_0 = np.mean(observed_0)
obs_treatment_1 = np.mean(observed_1)
```

Using pandas, the code can get even shorter:

```
shuffled = observation.copy() ①
random.shuffle(shuffled)
```

```
means = shuffled.groupby(treatment).mean() ②
means[1] - means[0] ③
```

① Here, we make use of the fact that observation is a column from a pandas dataframe. The function *random.shuffle* shuffles the values in place. To avoid overwriting the information in the original dataframe, we create a copy of the observations to preserve the original list.
② Both shuffled and treatment are pandas Series. The series (or list) that is used in the *groupby* function, doesn't need to be part of the series or dataframe that is grouped. The only condition is that they have the same length. Combining the *groupby* function with the *mean* function, we calculate the mean reduction for the treatment and control group.
③ The result of the previous statement is a pandas Series object, which allows us to subtract the two values to calculate the difference.

We are now ready to repeat this process 1000 times and collect the results in a list.

```
shuffled = observation.copy() # create a copy of the observations
differences = []
for _ in range(1000):
    random.shuffle(shuffled)  # shuffle the copy
    means = shuffled.groupby(treatment).mean()
    differences.append(means[1] - means[0])

print(f"Mean difference after reshuffling {np.mean(differences):.2f}")
print(f"Minimum difference {np.min(differences):.2f}")
print(f"Maximum difference {np.max(differences):.2f}")
```

Figure 4.3 shows the obtained distribution of the differences. It's created using the following code:

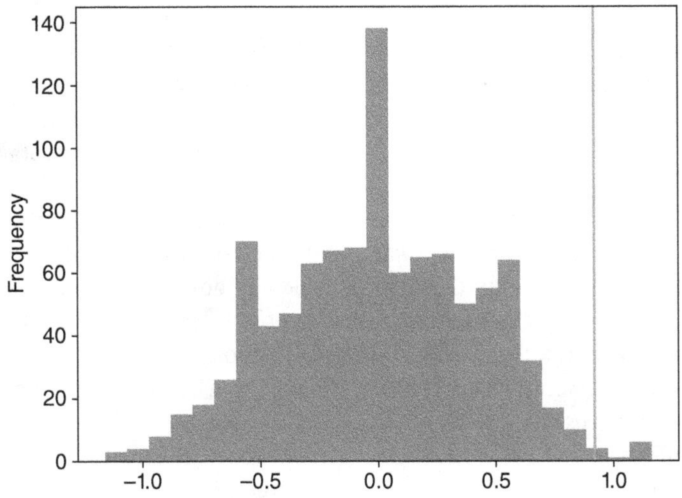

Figure 4.3 Histogram of 1000 trials—permutation of control and treatment groups (x-axis is the mean of the first shuffled group minus the mean of the second shuffled group).

```
ax = pd.Series(differences).plot.hist(bins=25) ①
ax.axvline(x=observed_difference, color="grey") ②
plt.show()
```

① We first convert the list into a `pandas` series and use its *Series.plot.hist* function to create the histogram plot.
② The *hist* function returns a `matplotlib` axis object (`ax`), which we can use to further customize the graph. Here, we add a vertical line at the observed difference.

Figure 4.3 clearly shows that the observed difference is very unlikely to occur by chance. We can calculate the probability of observing a difference of 0.92 or larger by chance by counting the number of times the difference is larger than 0.92 and dividing it by the total number of trials.

```
nr_greater_observed = sum(d >= observed_difference for d in differences) ①
prob_observed =  nr_greater_observed / len(differences)
print("Probability of observing a difference of 0.92 or larger by chance: "
      f"{prob_observed:.1%}")
```

① At first glance, this statement might look confusing. We iterate over all differences and for each difference `d`, we compare it to the observed difference. If it's larger, the comparison gives `True`, otherwise `False`. The *sum* function makes use of the fact that `True` is interpreted as 1 and `False` as 0. The value of the sum is therefore the number of differences that were larger than the observed difference.

Output

```
Probability of observing a difference of 0.92 or larger by chance: 1.1%
```

4.8.5 Write Functions for Code Reuse

In all our examples, we used functions from a variety of Python packages. Functions are an important concept in programming. They allow you to reuse code and make it easier to read and understand a program. Understanding how functions work and how to write them is an important skill for every data scientist.

Let's look at the resampling experiment again. While it was designed for the experiment with a specific dataset, we may want to run the same experiment with a different dataset. Instead of taking the original code and adjusting it, we can extract the core algorithm of the resampling experiment into a function and use that. Here is an example of how this can be done. The function takes two lists, the observations and the treatment groups, and the number of trials as arguments. It returns a list of differences.

The `def` keyword indicates that we are defining a function. The function name is `resampling_difference_means`. The arguments are `observations`, `treatments`, and `nr_trials`. The function body is indented and contains the code that is executed when the function is called. The function returns the list `differences`.

```
def resampling_difference_means(observations, treatments, nr_trials=1000): ①
    """ Calculate differences in means between two treatment groups
        using resampling """
    # create an independent copy of the observations
    shuffled = pd.Series(observations)
    differences = []
    for _ in range(nr_trials):
        random.shuffle(shuffled)  # shuffle the copy
        means = shuffled.groupby(treatments).mean()
        differences.append(means.iloc[1] - means.iloc[0])
    return differences
```

① The `nr_trials` is a so-called keyword argument with a default value of 1000. This means that we can call the function without specifying the number of trials. In this case, the default value is used. If we want to specify a different number of trials, we can do so by using the keyword argument.

An important aspect of this function is that we created an independent copy of the observations. If we would not create a copy, the original list would be shuffled after the function call. This is a so-called side effect, a common source of errors when writing functions. Look out for lists, dictionaries, and classes in the function arguments. These are mutable objects. It means that they can be modified in place. Numbers, strings, tuples, and booleans are examples of immutable objects. You can assign new values to them in a function without changing the value outside of the function. Sometimes, you want your function to have a side effect. The *random.shuffle* does exactly that. However, it is good practice to avoid side effects if possible.

Our function can now be used to repeat the resampling experiment with an increased number of trials.

```
differences = resampling_difference_means(df["Reduction"], df["Treatment"],
                                          nr_trials=2000)
print(f"Mean difference after reshuffling {np.mean(differences):.2f}")
print(f"Minimum difference {np.min(differences):.2f}")
print(f"Maximum difference {np.max(differences):.2f}")
```

Output

```
Mean difference after reshuffling -0.02
Minimum difference -1.24
Maximum difference 1.24
```

4.8.6 Organizing Code into Modules

In the previous section, we learned that functions help us to write reusable code that can be used in multiple projects. In fact, every time we import and use a function from a package like `pandas` we make use of *reusable* code. Once you start writing more functions and maybe even classes, you will find that you want to organize them into related groups. In Python, these groups are called modules.

It is good practice to keep your code organized in modules. You may even consider sharing your code with others by publishing it as a package.

Creating a module, a collection of Python functions and classes, is easy. You just create a file with the extension `.py` and add your functions and classes to it. For example, we could create a file `resampling.py` and put the function `resampling_difference_means` into it. We can then import the function using the `import` statement. This is what the file `resampling.py` could look like:

```
import random ①
import pandas as pd

def resampling_difference_means(observations, treatments,
                                nr_trials=1000):
    """ Calculate differences in means between two treatment
    groups using resampling """
    # calculate observed difference
    observed = observations.groupby(treatments).mean()
    # create an independent copy of the observations
    shuffled = pd.Series(observations)
    differences = []
    for _ in range(nr_trials):
        random.shuffle(shuffled)  # shuffle the copy
        means = shuffled.groupby(treatments).mean()
        differences.append(means.iloc[1] - means.iloc[0]) ②
    return {"observed": observed.iloc[1] - observed.iloc[0], ③
            "resamples": differences}
```

① We need to import all packages that are used in a module even if you already imported them somewhere else. Python is efficient and avoids reimporting the same package over and over again.

② We changed the index access from `[]` to `.iloc[]`. This is the recommended way to access elements of a pandas `Series` or `DataFrame`. The `[]` operator is ambiguous and can lead to unexpected results. If you change it to `means[1]-means[0]`, it will still work, but `pandas` will display a warning message that this will change in the future.

③ The function returns a dictionary with the observed difference and the list of differences from the resampling experiment.

If we want to use the function in a new project, we can import it using the `import` statement like every other module. For now, we assume that the file `resampling.py` is in the same directory as the file that we want to use it in. In this case, we write:

```
import numpy as np
import pandas as pd
import resampling ①

df = pd.read_csv("hybrid.csv")
differences = resampling.resampling_difference_means(df["Measurement"],
                                                      df["Experiment"]) ②

resamples = differences["resamples"]
```

```
print(f"Observed difference {differences['observed']:.2f}")
print(f"Mean difference after reshuffling {np.mean(resamples):.2f}")
print(f"Minimum difference {np.min(resamples):.2f}")
print(f"Maximum difference {np.max(resamples):.2f}")
```

① Import of our module by name.
② Because we import the module by name, we need to prefix the function name with the module name.

Output

```
Observed difference -240.59
Mean difference after reshuffling -1.61
Minimum difference -141.47
Maximum difference 125.47
```

In this example, we imported the full module. Alternatively, we could have only imported the function `resampling_difference_means` using the `from` statement.

```
from resampling import resampling_difference_means
```

This allows us to use the function without prefixing it with the module name. You will also come across *wildcard* imports, which import all functions from a module. For example:

```
from resampling import *
```

This is not recommended because it can lead to name clashes. This means, if we import functions with the same name from different modules, the second import will overwrite the first. There will be no warning and it can be hard to find the source of the problem. It is always better to write explicit imports and reduce the imports to what is needed.

Exercises

4.1 DISCUSS: What are the advantages and disadvantages of setting the voting age very high? Or very low?

4.2 Answer the following questions, referring to the *pulse.csv* data, and your answers to problems with the same data in Chapter 2. Use Python to perform the calculations.

a) Is there any sign that the proportion who ran is different for males vs. females? Perform a hypothesis test and state your conclusion in plain language that would be understandable to someone who never took a statistics course.

b) Is there evidence that the "before" pulse rate differs between smokers and nonsmokers? Medical theory suggests that smoking elevates the pulse rate. Perform a hypothesis test.
c) We would expect running in place to affect a person's pulse rate. Test whether before and after pulse rates are the same.

4.3 Each of the following actions introduces the risk of a Type I error or a Type II error. For each case, which type of error lurks as a risk? Also discuss the likely cost of the error.
a) A new diet pill is introduced to the market after passing an appropriate statistical review by the drug regulatory agency.
b) A computer chip manufacturer tests a new process and the results show it produces faster chips. However, the improvement is not statistically significant, so the process is not developed further.
c) A municipal water supplier tests water regularly, and a particular test shows a high level of contaminants. The water supplier issues an alert for residents not to drink or cook with the water.
d) A parks department tests water in a lake to determine the level of fecal coliform bacteria, which can pose risks of gastrointestinal illness for swimmers. A particular test yields a high level of coliforms, but due to some uncertainty about the reliability of the lab results no action is taken.

4.4 A web consultant for a retail merchant tests two different presentations of a product to determine whether there is a difference in the length of time that people stay on that product page. She records the visit time in seconds per visitor for each presentation and finds that people viewing Presentation B (the new one) linger 0.33 seconds longer, on average, than those viewing Presentation A (the existing presentation). In the time allotted for the study, she is able to record 36 visits, 21 for Presentation A and 15 for Presentation B.
a) From the histogram shown in Figure 4.4 for a permutation test between the two treatments (putting all 36 values in a hat, repeatedly shuffling and drawing out resamples of 21 and 15 values for A and B respectively, and recording the difference), is the observed difference within the range of chance variation? You can load the dataset from *web-page-data.csv* and perform the resampling experiment yourself.
b) Now suppose that Presentation A and Presentation B are both part of a web remake, and neither is an established standard. How would your statistical approach differ?

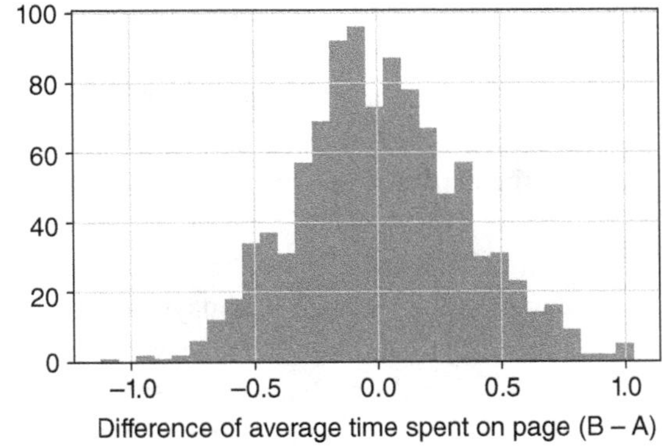

Figure 4.4 Histogram of 1000 resampled difference in the length of time people stay on the product page for presentations B compared to A.

4.5 From Figure 4.2, the Normal curve, estimate the proportion of values that are:

a) Greater than +2 standard deviations

b) Less than −2 standard deviations

Use `scipy.stats.norm.cdf` to check your answers.

4.6 Implement the following dice roll experiments using Python:

a) Roll a dice 100 times and count the number of times you roll a 6 (or any other number). What count do you expect? Implement this experiment using Python and compare the result to your expectation.

b) Roll two dice 100 times and count the number of times you roll a 7 (or any other number).

c) Roll two dice 100 times and count the number of times you roll a 7 (or any other number). Repeat this experiment 1000 times and plot the distribution of the number of times you roll a 7.

4.7 In 2023, cracking a 10-character password that contains only lowercase letters takes minutes. In order to make passwords more secure, a company decides to require that all passwords contain at least one uppercase letter, one lowercase letter, one number, and one special character. The company also requires that all passwords be at least 12 characters long. Write a Python program that generates a random password that meets these conditions:

- at least 12 characters long
- contains at least one uppercase letter
- contains at least one lowercase letter
- contains at least one number
- contains at least one special character

4.8 Toss a coin 1000 times and count the length of sequences of consecutive heads or tails. This is called a run. Plot the distribution of the run lengths. What distribution do you expect?

4.9 Birthday paradox: How many people do you need in a room to have a 50% chance that two people have the same birthday? How many people do you need in a room to have a 99% chance that two people have the same birthday? Assume that all birthdays are equally likely and ignore leap years. Write a Python program to answer these questions.

a) Simulate the problem by generating random birthdays until you see one for the second time. Store the number of birthdays you had to generate. Repeat this process 10,000 times and store the number of birthdays for each trial in a list.

b) Count how often each number occurs and convert it into a frequency table. Convert the counts into probabilities and add them to the table together with a cumulative probability.
Hint: You can use a `defaultdict` to count the occurrences.

c) Plot the distribution of the number of birthdays you had to generate.

d) Use the cumulative property to answer the questions.

4.10 Determine the number π using integration with a Monte Carlo simulation. Draw a square with side length 2 and a circle with diameter 2. Generate random points in the square and count the number of points that are also in the circle. The ratio of the number of points in the circle to the total number of points is an approximation of $\pi/4$. Repeat this experiment 20 times with an increasing number of points and plot the approximation of π as a function of the number of points for 10,000, 20,000, …, 100,000. Describe your findings.

4.11 In this exercise, we look at the effect of using n or $n - 1$ in the denominator when calculating the standard deviation of a sample.

a) Create a random sample of 1000 values from a Normal distribution with mean 100 and standard deviation 15.

b) Compute the standard deviation of these 1000 values.

c) Repeatedly, take resamples of size 10 and calculate the standard deviation of each resample using either n (population) or $n - 1$ (sample) in the denominator.

d) Plot the distribution of these standard deviations for both cases and compare to the standard deviation of the original 1000 values.

e) What do you observe?

4.12 This problem uses the dataset *trade-discount-A-B.csv* (There are similar problems in Chapters 2 and 11). Consumer packaged goods (CPG) companies sell some goods online but generate most of their revenue from sales through brick and mortar retail companies. A CPG company typically manages a number of brands, each of which has multiple products. For example, Unilever, a large CPG based in the United Kingdom, owns over 400 brands. Just one of its brands, Dove, sells body washes, hand and body lotions, facial cleansers, deodorants, shampoos, conditioners, and hair styling products. Each product is sold in multiple sizes and variations (e.g. different soap scents), so the number of individual "stockkeeping units" (SKU's) on offer from a CPG might be in the tens of thousands. A typical grocery or drug store has room to stock only a small fraction of all the products on offer from a given CPG. A major challenge for the CPG is obtaining and retaining "shelf space" at major retailers. CPG companies promote their products through advertising, issuance of discount coupons to consumers, and "trade discounts" offered to retailers. Trade discounts are product-specific discounts offered to individual retailers that can fund co-advertising, or simply serve as an incentive to the retailer to promote the discounted product. The company is debating whether to reduce trade discounts and boost coupons, as is done in some of its Latin American markets. Analysts are considering whether and how to conduct an experiment (A/B test) to shed light on whether this is a good idea.

a) Many in the company resist the idea of an experiment, as it would be complex and costly, disrupting many relationships with retailers. Moreover, if the company were to change the trade discounting policy and observe what happened to sales during the experiment period, there might be other changes that affect sales and muddy the picture. In the end, it is agreed to select two mid-sized US markets that have a similar mix of product sales. In one of the markets, trade will be reduced and coupon spend increased. No changes will be made in the other (control) market. Could a randomized experiment be used here instead? Why or why not?

b) The data show the A/B test results, where C = the control group where no changes to current practice have been made, and T = treatment group, where trade discounts have been boosted and coupon spending reduced, so as to leave overall promotion spending approximately unchanged. Report the overall result, and its statistical significance. Interpret the business significance of the result as well.

5

Probability

"Is this a game of chance?"
"Not the way I play it." W. C. Fields
Source: Courtesy of Universal Studios Licensing LLC.

We have been dealing regularly with the notion of probability; in this chapter we introduce more formal concepts of probability. After completing this chapter, you should be able to:

- produce a Venn diagram
- use the addition rule and explain in what circumstances it is relevant
- explain the binomial probability distribution
- calculate standardized values (z scores)
- interpret the role of the Normal distribution as a benchmark

5.1 What Is Probability

Most people consider that they have an intuitive sense of probability and an idea of chance. The weather forecaster does not need to explain what is meant by "chance

Statistics for Data Science and Analytics, First Edition. Peter C. Bruce, Peter Gedeck, and Janet Dobbins.

Companion website: www.wiley.com/go/Wiley_Statistics_for_Data

of rain." Formal and scientific understandings are more elusive—volumes have been written on probability over the centuries, in the realms of history, philosophy, and mathematics. So we will not get tangled up in formal definitions of probability. Instead, we will point to two useful concepts in interpreting probability:

1) Long-run Frequency: Probability can be seen as the frequency with which an outcome of the event would occur if repeated over and over. For example, the proportion of time you would get "heads" if you flip a coin many times. This is easiest to understand for a concrete process, such as a game of chance, whose repetition is easy to visualize. The earliest expositions on the theory of probability were aimed at helping gamblers better understand the odds (closely related to probabilities) in games they were playing.
2) Degree of Belief: Probability can be seen as a numerical mapping of the degree to which you believe something will occur. For example, military planners might attach a probability to the outcome, "Pakistan and India will fight a war in the next five years." It is difficult to imagine a repeatable process in which this question can be framed, but the lack of such a practical process does not diminish the relevance or utility of the concept of probability as applied to the question.

5.2 Simple Probability

Perhaps no one is more familiar with probability in its pure form than gamblers. The 17th century writer Antoine Gombaud, better known under his assumed moniker Chevalier de Mere, was a prolific gambler at dice. The foundations of probability theory were laid when he enlisted the help of mathematician Blaise Pascal (shown in Figure 5.1) to explain why he was consistently losing his bets on getting two sixes in 24 throws of a pair of dice. He assumed the probability was the same as throwing one six in 4 plays of a single die, where he consistently won his bets.

Pascal showed this simple calculation to be wrong and, in his analysis, showed the importance of enumerating all the possible outcomes of throwing a pair of dice 24 times.

Definition: Sample Space The list of all possible outcomes of a specified experiment or event is called the sample space.

The sample space for a coin toss is heads, tails. The sample space for the throw of a die is 1, 2, 3, 4, 5, 6. The sample space for "flight arrival status" could be early, on-time, delayed, or canceled. In the flight arrival case, as with most cases in real

Figure 5.1 Blaise Pascal, 1623–1662. Source: Unknown author/Wikimedia Commons/CC BY 3.0.

life, the sample space may not fully flow from deterministic rules as it would in a gambling game. For example, defining "on-time" for a flight may include a buffer in which the flight might be a few minutes late.

The sample space must:

- include every possible outcome,
- be *jointly exhaustive*, i.e. as a whole include every possible outcome, and
- *mutually exclusive*, i.e. the outcomes on the list must not overlap.

Sample space is a statistical concept, but the idea is also important in software engineering. Incomplete or inaccurate specification of possible outcomes can cause bugs, if a user enters data that is not part of the anticipated states or outcomes. For example, American Airlines' website is supposed to incorporate the official code for each destination it serves, to facilitate online purchases. Throughout 2023, a customer entering the code MVY (for Martha's Vineyard, an American destination) would encounter a fatal error at the end of the purchase process that crashed the site. The problem? The set of valid destinations did not include MVY.

Subsets of the sample space are often of interest—for example, the flight arrival outcomes "delayed or canceled." A weather outcome of special interest might be "freezing rain or snow."

A list of possible outcomes might not be the same thing as the sample space. A text classification algorithm might classify aircraft maintenance tickets as "urgent," "not urgent," "safety-related," or "not safety-related." An item will receive one label from the urgency category and one from the safety category, and both are needed. An item for an aircraft in service might be urgent but not safety-related (e.g. a malfunctioning toilet). An item, say an engine fault for a backup aircraft not in service, might be nonurgent and safety related. A full description of the sample space requires a set of four *compound* outcomes:

- Urgent but not safety related
- Urgent and safety related
- Nonurgent and not safety related
- Nonurgent and safety related

Unlike the initial lists, these compound outcomes are mutually exclusive and exhaustive, and illustrate that many scenarios require analysis and processing to arrive at an accurate description of the sample space.

5.2.1 Venn Diagrams

It is common to draw pictures called Venn diagrams for simple probability situations. In Figure 5.2, the disk represents event *E* (for example, rain) and the gray area represents the "complement" ~*E* (not rain). The entire rectangle represents the whole sample space with an area (probability) of 1.

(?) The probability of an event plus the probability of its complement add up to ___?

Figure 5.3 depicts a situation where two outcomes overlap. For today's weather, there may be a certain probability of rain (rectangle *B*), a certain probability of snow (rectangle *E*), a certain probability of both rain and snow (the intersection of *B* and *E*), and a certain probability of neither rain nor snow (the gray area outside *B* and *E*). If no outcomes occur together, the Venn diagram would show no overlapping area. In calculating probabilities from the Venn diagram and the outcomes it represents,

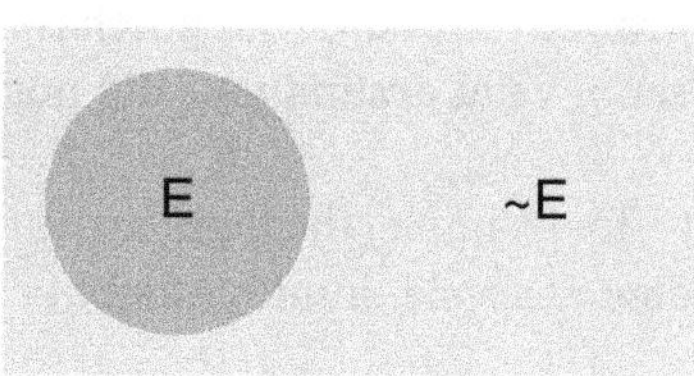

Figure 5.2 Venn diagram: E and ~E (~E means "not E").

Figure 5.3 Venn diagram—two events of interest (B and E) and their intersection (B∩E).

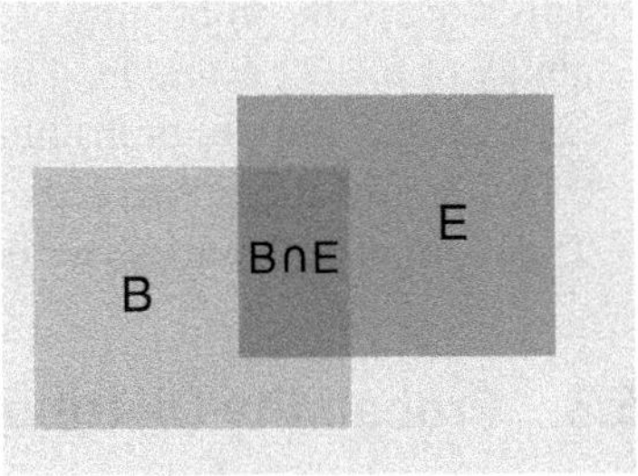

1) The probabilities of all the outcomes must sum to 1
2) If outcomes *A* and *B* are mutually exclusive (non-overlapping in the Venn diagram), the probability of *A* OR *B* is the sum of the two probabilities
3) If outcomes *A* and *B* overlap, the probability of *A* OR *B* is the sum of the two probabilities MINUS the probability of the two happening together (the intersection)
4) The probability of *A* AND *B* is the probability associated with the intersection region.

In programming logic the operator *AND* is used where two or more outcomes *both* occur: "list all the customers who purchased product A and product B." For a customer to be listed, they must have purchased *both* products.

Don't let your intuition lead you astray with the usage of AND and OR. "AND" makes us think of including both B outcomes and A outcomes, but we really want to include only those events that have both outcomes A and B. "OR" makes us think of having to choose one outcome or another, but we really want to include all *events* that have either A or B as an outcome.

Try It Yourself

Make a table showing the sample space for three tosses of a coin. *Hint:* Use the table for two tosses as a starting point. Then note that each item on that list has two children. For example, the parent TH has children THT and THH.

The following rule is a simplified version of the Venn diagram, and it applies only for mutually exclusive outcomes to an event:

Addition Rule: $P(\mathrm{A} \cup \mathrm{B})$

For *mutually exclusive* (disjoint) outcomes:

$$P(\mathrm{A} \cup \mathrm{B}) = P(\mathrm{A}) + P(\mathrm{B})$$

This reads, "the probability of A or B equals the probability of A plus the probability of B." For example, if the probability of the Yankees winning the World Series is 0.2 and the probability of the Nationals winning the World Series is 0.15, the probability that either the Yankees OR the Nationals will win is 0.35. They cannot both win, so we do not need to subtract out that probability.

5.3 Probability Distributions

In Chapter 3, we were introduced to frequency distributions and histograms to sum up and show how a data set is distributed. Let's now apply frequency distributions to probabilities. For example, here we tabulate estimated probabilities associated with flights in September, 2023 between two US airports, Atlanta (the busiest) and NY-Newark (the most-delayed):

On-time	0.24
Delayed	0.70
Canceled	0.06

These outcomes are mutually exclusive (the categories do not overlap), and their probabilities sum to 1.

5.3.1 Binomial Distribution

An important probability distribution in statistics and data science is the binomial (two-outcome) distribution. It is a widely used concept because so many scenarios that data scientists study are binomial outcomes—buy or don't buy, fraud or no-fraud, click on link or no click, survive or die. Many situations are more complicated, of course, but, even so, it is typically the case that a decision is needed, and it often boils down to a yes–no decision.

A binomial distribution shows the frequencies of obtaining x successes in n trials, where each trial has a p probability of success. "Success" is an arbitrary term simply denoting the outcome we choose to be interested in. A "trial" is whatever we define as the yes–no event of interest (for example, a coin flip, examination of one insurance claim to determine whether it is fraudulent, showing a web visitor a page with a clickable link, a pitch to a batter in baseball). Here are those elements for the Try it Yourself flipping a coin three times experiment (above):

- Trial: flipping a coin
- Success: heads
- Probability of success: 50%
- Number of trials: 3

"Success" here is a purely statistical term and does not imply a favorable outcome. In cases where we are interested in relatively infrequent outcomes (a fraudulent insurance claim, a patient's death), "success" typically denotes the rare outcome of interest, though it may be quite unfavorable in practical terms.

Try It Yourself

Using your table showing the sample space for three tosses of a coin, add probability calculations for each possible outcome. Assume the coin has a 50/50 chance of landing heads or tails—that it is a "fair coin." Once you have your table, find the probability of getting more heads than tails.

We can easily adapt our simple example to cover a wide variety of situations by modifying the coin so it is "unfair." For example, if we want to model a 2% probability that a web visitor will click on a link, we would use a (theoretical) coin that has a 2% chance of landing heads. More practically, we would use random numbers. If we randomly select integers between 1 and 100, the numbers "1" and "2" could be "click" and the remainder "no-click."

Enumerating all possible outcomes for our simple experiment with three flips of a coin was relatively easy. This becomes prohibitively hard for large *n* problems. There are three possible solutions:

1) We can simulate probabilities with repeated resampling
2) We can calculate probabilities with the binomial formula
3) We can estimate probabilities using the Normal distribution

The binomial formula and the use of the Normal distribution to estimate binomial probabilities are found in the Appendix to this chapter.

Here is the Python code for repeated resampling of 150 flips of a fair coin. The resulting distribution is shown in Figure 5.4.

```
import random
import pandas as pd

random.seed(123) ①
nrepeat = 10_000
nr_heads = []
for _ in range(nrepeat):
    nr_head = random.choices(["H", "T"], k=150).count("H") ②
    nr_heads.append(nr_head) ③

ax = pd.Series(nr_heads).plot.hist(bins=np.arange(49.5, 100, 1)) ④
ax.axvline(75, color="black", linestyle="-")
ax.set_xlabel("Number of heads")
```

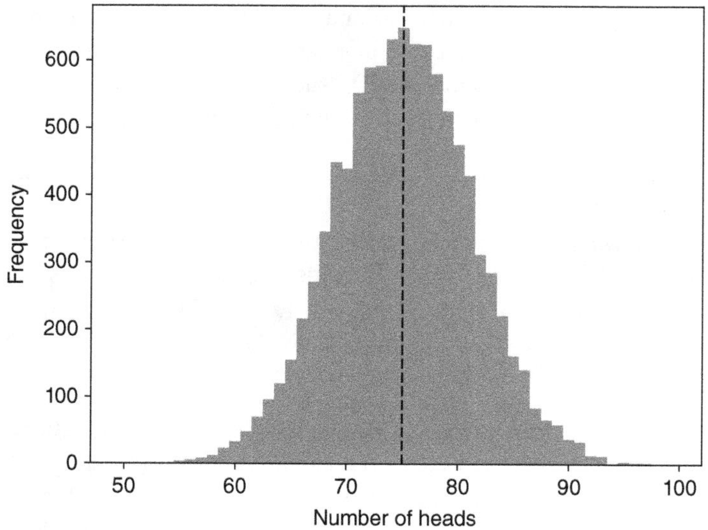

Figure 5.4 Distribution of the number of heads in 150 flips of a fair coin.

① Set the random number generator seed so that the results are reproducible
② Randomly select 150 values from the list ["H", "T"] with replacement and count the number of "H" values
③ We could replace the `for` loop with a list comprehension:
```
nr_heads = [random.choices(["H", "T"], k=150).count("H") for _
in range(nrepeat)]
```
④ We convert the list to a `Pandas` Series to use its plot function

5.3.1.1 Example

A baseball batter has a 0.3 probability of getting a hit in each at-bat. What is the probability that he will get exactly three hits in five at-bats?

The binomial distribution shows the probability of all possible outcomes for a given probability and number of trials. The distribution shown in Table 5.1 and Figure 5.5 shows the distribution of success probabilities ("success" = hit) when $n = 5$ and $p = 0.3$. It can be estimated by resampling and calculated by the binomial formula (see Appendix to this chapter).

? Answer the following questions based on the information in this table:

1) What is the probability of getting four successes in 5 trials?
2) What is the probability of getting two or fewer successes?
3) What series of steps (algorithm) would you use to create the above table using resampling?

Table 5.1 Probability of different successes.

Successes	Probability
0	0.168
1	0.36
2	0.309
3	0.132
4	0.028
5	0.002

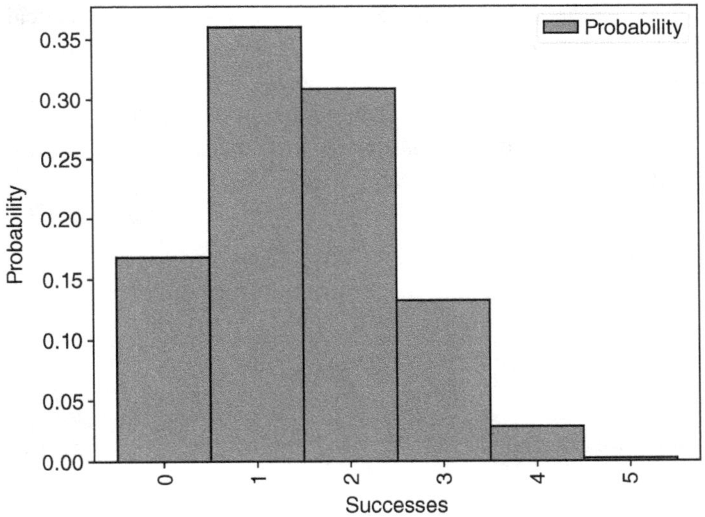

Figure 5.5 Probability of successes (hits) in 5 at-bats, calculated using the binomial formula and estimated using resampling.

5.4 From Binomial to Normal Distribution

Go back and look at Figure 5.4. The shape of this distribution may be familiar to you—it is the bell-shaped form of the Normal distribution the we introduced in the last chapter (see Figure 5.2). The Normal distribution has been central to statistics for well over a century and is used in many circumstances, including:

- To approximate the binomial distribution when n is large and
- To approximate the permutation (shuffling) distribution that we explored in earlier chapters

The use of approximations like the Normal curve and its cousin, the t-distribution, were the standard approach in statistics until the arrival of widely available computing power that allows data scientists to directly model problems with appropriate simulations.

5.4.1 Standardization (Normalization)

Much of our work to this point has been to determine how improbable (extreme) an observed value is, relative to a resampling distribution of values under the null hypothesis or chance model. Before computers were available, producing that chance distribution was impractical—it would require lengthy sessions dealing cards from a box or tossing dice.

Using a mathematical approximation, though, hits a roadblock. It is impractical to derive a different mathematical benchmark each time you want to conduct a study, to accommodate the scale of your measurements. One study might have data centered around 15 meters with a standard deviation of 3 meters, another with a mean of 25 μm with a standard deviation of 2 μm. Instead, we *standardize* our data so that all distributions are on the same scale.

Definition: Standardization and z-Scores We standardize or normalize values in a sample or dataset by subtracting the mean from each and then dividing by the standard deviation.

$$\frac{x_i - \mu}{\sigma}$$

Standardized values are also called z-scores.

Standardizing data in this way strips scale and units from the measurement. For example, instead of saying that a person weighs 220 pounds, we would say that they weigh 1.8 standard deviations above the mean. You will also encounter the term *normalizing*, which means the same thing. The term "normalizing" is not directly related to the Normal distribution—normalizing data does *not* make it have a Normal distribution. It does, however, make it easy to compare your data to a standard Normal distribution (see Section 5.4.2).

Table 5.2 illustrates the calculations for cholesterol scores for a group of 10 subjects. For this group, the mean cholesterol is 204.9 and the standard deviation is 22.81. Consider subject #6.

- Cholesterol raw score: 224
- Standardized score: $\frac{224 - 204.9}{22.81} = 0.837$

Table 5.2 Cholesterol scores for a group of 10 subjects.

Subject	Cholesterol
1	175
2	210
3	245
4	198
5	210
6	224
7	189
8	171
9	232
10	195
Mean	204.90
SD	22.81

So the standardized cholesterol score for subject #6 is 0.837. Another way of saying this is that subject six's cholesterol is 0.837 standard deviations above the average.

5.4.2 Standard Normal Distribution

The single Normal distribution that is used as a benchmark is called the Standard Normal distribution.

Definition: The Standard Normal Distribution The Standard Normal distribution has a mean of 0 and a standard deviation of 1.

Standard Normal distribution graphs provide standardized values (z-scores) on the x-axis. The y-axis is labeled "density," which is not meaningful by itself. What is meaningful is the area *under* the curve. The total area is = 1, and the area between or beyond points on the x-axis is the probability that x takes on a value between (or beyond) those points. For example, the shaded area in Figure 5.6 is the probability that x is greater than 1.

Calculating this probability from the graph is not easy, so cumulative probability tables, z-tables, are typically used.

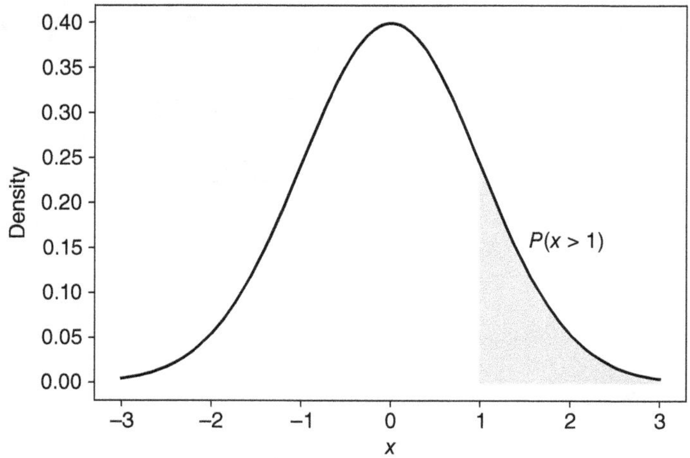

Figure 5.6 $P(x > 1)$.

5.4.2.1 *z*-Tables

Z-tables, also called Standard Normal tables, show the probability that a value has a value lower than the specified *z*-score in the table. Table 5.3 is a very simple table. If a dataset follows the Normal distribution and has been standardized (normalized), the first row of numbers in the table says the probability that a value z will be less than −3 is 0.001350. So although any number less than −3 is possible, such numbers are not very likely.

The last row of the table says that the probability that z is less than +3 is 0.998650, which is pretty likely. Since the cumulative probability must equal 1, we also know that the probability that z is more than 3 is $1 - 0.998650 = 0.001350$. The Normal distribution is symmetrical, so the probability of being more than 3 is the same as the probability of being less than −3.

Table 5.3 Simple *z*-table.

z	$P(Z < z)$
−3	0.001350
−2	0.022750
−1	0.158655
0	0.500000
1	0.841345
2	0.977250
3	0.998650

Older statistics textbooks contain detailed versions of the above table. It is more common, though, to find the probability values directly in statistical software, or via a web calculator (do a web search for "*z*-score calculator").

> **Try It Yourself**
> For subject number 8 in the cholesterol table above (Table 5.2), find (1) the *z*-score, and, using software or a web-based calculator, (2) the cumulative probability associated with that score. Then interpret that probability.

5.4.3 The 95 Percent Rule

We can see from Table 5.3 to find that the probability that a Standard Normal variable is above 2 or below -2 (in other words, beyond two standard deviations from the mean) is $0.02275 + 0.02275 = 0.04550$. A guideline based on the calculation we just did is that about 95% of any Normal distribution is within two standard deviations of the mean.

> **Try It Yourself**
> What percentage of the standard Normal distribution is within one standard deviation of the mean? What about three standard deviations?

These guidelines are widely used for detection of *outliers*, which are points unusually distant from the mean and therefore worthy of special attention. Rules like this for finding outliers should be used with caution. Just because an observation is distant from most of the others does not necessarily mean it is anomalous in other respects.

The 95% rule is valid only for Normally distributed data. Even minor deviations from Normality can have a big impact at the tails of the distribution.

5.5 Appendix: Binomial Formula and Normal Approximation

The *binomial formula* tells you the probability of getting exactly x successes in n trials, when the probability of success on each trial is p. The formula is:

$$\begin{aligned} P(X) &= \binom{n}{x} p^x (1-p)^{n-x} \\ &= \frac{n!}{x!(n-x)!} p^x (1-p)^{n-x} \end{aligned}$$

5.5.1 Normal Approximation

The binomial formula can get cumbersome to calculate, especially when calculating cumulative probabilities that require many calculations of individual probabilities. When n is large, we can take advantage of the fact that the Normal distribution looks almost exactly like the binomial distribution, and use z-scores. The sample proportion has a standard error of $(p(1-p)/n)^{1/2}$, so we can divide by this to obtain z-scores.

For example, suppose 100 customers are surveyed and 60% of them have a favorable opinion of your product. But could opinion among all customers be equally split, and, due to sampling variation, you got a set of 100 that was 60% favorable? Specifically, you'd like to know the probability that a 60% sample could come from a 50% population. This is a binomial problem in proportions, so we can divide the difference between observed (0.60) and hypothesized (0.50) = 0.10, and divide by the standard error calculated from the sample:

$$\frac{0.1}{\sqrt{p(1-p)/100}} = \frac{0.1}{\sqrt{0.5(1-0.5)/100}}$$
$$= 0.1/0.05$$
$$= 2$$

In other words, two standard deviations separate 60% from the hypothesized 50%. We saw earlier from the z-table that the probability of a z-score of 2 or more is small—0.2275.

5.6 Python: Probability

5.6.1 Converting Counts to Probabilities

We often want to convert counts to proportions or estimated probabilities (or, equivalently, frequencies to relative frequencies). This is easy to do in Python. You divide each count by the sum of all counts. In the following example, we throw a die 20 times and record our results. This code illustrates this for a list of counts stored in the dictionary `counts`:

```
import random
random.seed(1245)

counts = {1: 0, 2: 0, 3: 0, 4: 0, 5: 0, 6: 0} ①
for _ in range(20):
    outcome = random.randint(1, 6) ②
    counts[outcome] += 1 ③
sum_counts = sum(counts.values()) ④
probabilities = {key: count / sum_counts for key, count in counts.items()} ⑤
```

```
print(f"counts: {counts}")
print(f"sum of counts: {sum_counts}")
print(f"probabilities: {probabilities}")
```

① counts is a dictionary that stores the counts for each outcome. It is initialized to zero for each possible result from 1 to 6.
② random.randint(1, 6) returns a random integer between 1 and 6, inclusive.
③ counts[outcome] += 1 increments the count for the observed outcome by 1.
④ sum(counts.values()) returns the sum of the counts. The dictionary method values iterates over the values in the dictionary. In this case, the values are the counts.

Output

```
counts: {1: 2, 2: 6, 3: 3, 4: 2, 5: 5, 6: 2}
sum of counts: 20
probabilities: {1: 0.1, 2: 0.3, 3: 0.15, 4: 0.1, 5: 0.25, 6: 0.1}
```

The idea is the same, independent of how your data are stored. The following demonstrates it for a list of counts:

```
counts_list = [2, 6, 3, 2, 5, 2]
sum_counts = sum(counts_list)
probabilities = [count / sum_counts for count in counts_list]
```

It is also straightforward if the data are stored in pandas data structures.

```
import pandas as pd
counts_df = pd.DataFrame({"outcome": k, "count": v}
                         for k, v in counts.items()) ①
counts_df["probability"] = (counts_df["count"] /
                            counts_df["count"].sum()) ②
counts_df
```

① The *items* method returns the key-value pairs in the dictionary. We convert each pair to a dictionary and pass this sequence to the constructor of the *DataFrame*. This will create a dataframe with two columns: outcome and count.
② This statement divides each value in the count column by the sum of the values in that column. The result is stored as a new column in the dataframe.

Output

```
   outcome  count  probability
0        1      2         0.10
1        2      6         0.30
2        3      3         0.15
3        4      2         0.10
4        5      5         0.25
5        6      2         0.10
```

5.6.2 Probability Distributions in Python

In Python, we have many options to create random numbers for different probability distributions. The random package implements the most frequently used distributions and for many applications this will be sufficient. If you need a distribution that is not supported by `random`, check `numpy` or even better `scipy`. The `scipy` package implements a wide variety of distributions and provides many useful functions for working with them.

5.6.3 Probability Distributions in `random`

Here are a few examples that demonstrate how we can generate random numbers from different distributions using `random`.

```
import random
random.seed(321)

uniform = [random.uniform(3, 10) for _ in range(3)] ①
normal = [random.gauss(1, 3) for _ in range(3)] ②
poisson = [random.expovariate(0.5) for _ in range(3)] ③

def print_numbers(name, numbers): ④
    s = ", ".join([f'{number:.3f}' for number in numbers])
    print(f"{name}: {s}")

print_numbers("uniform", uniform)
print_numbers("normal", normal)
print_numbers("poisson", poisson)
```

① *random.uniform(a, b)* returns a random number from a uniform distribution between a and b.
② *random.gauss(mu, sigma)* returns a random number from a Normal distribution with mean `mu` and standard deviation `sigma`.
③ *random.expovariate(lambd)* returns a random number from an exponential distribution with rate `lambd`. The argument is called *lambd* because `lambda` is a reserved word in Python.
④ We created the function to make the code more readable.

Output

```
uniform: 4.928, 3.879, 6.932
normal: -6.217, 3.752, 7.947
poisson: 1.214, 2.938, 2.187
```

Looking at the numbers doesn't tell us much. Let's create a figure that compares the three distributions. Figure 5.7 shows the result of the following code. The distribution of the generated random numbers are clearly different.

```
fig, axes = plt.subplots(figsize=(10, 2.5), ncols=3)
axes[0].hist([random.uniform(3, 10) for _ in range(5000)],
             bins=50, density=True)
axes[0].set_title("uniform")
axes[1].hist([random.gauss(1, 3) for _ in range(5000)],
             bins=50, density=True)
axes[1].set_title("normal")
```

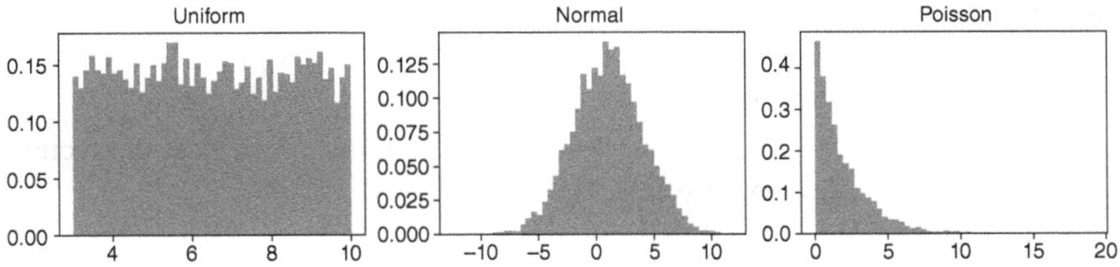

Figure 5.7 Histograms of 5000 random numbers sampled from different distributions.

```
axes[2].hist([random.expovariate(0.5) for _ in range(5000)],
             bins=50, density=True)
axes[2].set_title("poisson")
plt.tight_layout()
```

5.6.4 Probability Distributions in the `scipy` Package

5.6.4.1 Continuous Distributions

The `scipy` package implements more than 100 distributions. As an example, we look at the Normal distribution. The following code generates 5000 random numbers from a Normal distribution with mean 1 and standard deviation 3.

```
import numpy as np
from scipy import stats

rng = np.random.default_rng(seed=783)
samples = stats.norm.rvs(loc=1, scale=3, size=5000, random_state=rng)
print(samples)
```

Output

```
[3.87280849  3.49605883  1.55079719 ... -1.93997238 -2.76834212  5.73210174]
```

The function *stats.norm.rvs* takes a variety of arguments. The first two arguments are the mean (`loc`) and standard deviation (`scale`) of the distribution. The argument `size` specifies the number of random numbers to generate. Finally, we set the optional `random_state` for reproducibility.

There is an alternative way of using the Normal distribution in `scipy`. In the following example, we create a `dist` object that represents a so-called frozen distribution (mean `loc` and standard deviation `scale` are fixed). When we use this object, we no longer need to specify the mean and standard deviation. The following code illustrates this. This approach is useful if you want to work with a distribution where the key properties are fixed.

```
rng = np.random.default_rng(seed=783)
dist = stats.norm(loc=1, scale=3)
samples = dist.rvs(size=5000, random_state=rng)
```

Output

```
[3.87280849  3.49605883  1.55079719 ... -1.93997238 -2.76834212
   5.73210174]
```

We can also use the `stats.norm` object and functions to calculate characteristics of the distribution. For example,

```
dist = stats.norm()  # standard normal distribution
print("mean:", dist.mean())
print("standard deviation:", dist.std())
print("median:", dist.median())
print("variance:", dist.var())
print("(mean, variance, skewness, kurtosis)", dist.stats(moments="mvsk"))
```

Output

```
mean: 0.0
standard deviation: 1.0
median: 0.0
variance: 1.0
(mean, variance, skewness, kurtosis) (0.0, 1.0, 0.0, 0.0)
```

A convenient function is also the *interval* function. It returns the interval that contains a given percentage of the distribution. For example, the following code returns the intervals that contains 90% and 95% of the distribution.

```
print("90% interval:", dist.interval(0.9))
print("95% interval:", dist.interval(0.95))
```

Output

```
90% interval: (-1.6448536269514729, 1.6448536269514722)
95% interval: (-1.959963984540054, 1.959963984540054)
```

Figure 5.8 shows the graphs of five functions that describe the probability distribution.

***pdf*:** The probability density function (pdf) describes the probability of a random variable taking on a specific value.

***cdf*:** The cumulative distribution function (cdf) describes the probability of a random variable taking on a value less than or equal to a specific value. For example, `dist.cdf(1.645)` is approximately 0.95. This means that 95% of the distribution is less than or equal to 1.645.

***sf*:** The survival function (sf) is the complement of the cdf. For example, `dist.sf(1.645)` is approximately 0.05. This means that 5% of the distribution is greater than 1.645.

***ppf and isf*:** The percent point function (ppf) is the inverse of the cumulative density function and the inverse survival function (isf) is the inverse of the survival function.

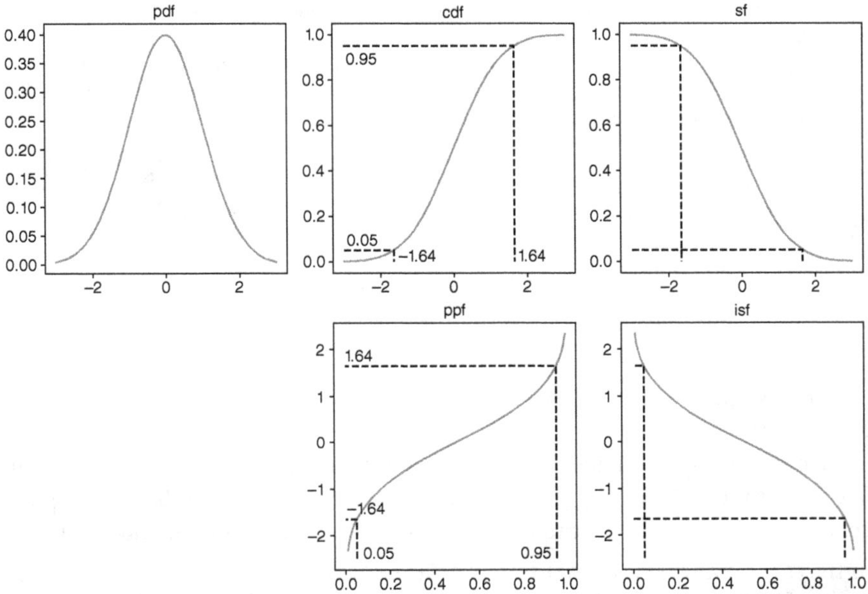

Figure 5.8 Probability density function (pdf), cumulative distribution function (cdf), survival function (sf = 1-cdf), percent point function (ppf), and inverse survival function (isf) for the standard normal distribution.

Using the *cdf* function, we can confirm the "95 Percent rule."

```
within_1std = stats.norm.cdf(1) - stats.norm.cdf(-1)
within_2std = stats.norm.cdf(2) - stats.norm.cdf(-2)
within_3std = stats.norm.cdf(3) - stats.norm.cdf(-3)
print(f"within 1 std of mean: {within_1std:.2%}")
print(f"within 2 std of mean: {within_2std:.2%}")
print(f"within 3 std of mean: {within_3std:.2%}")
```

5.6.4.2 Discrete Distributions

Discrete distributions are handled in a similar way. The following example shows how to work with the binomial distribution that we used for the baseball batter example from Section 5.3. We create a frozen distribution object and then use it to calculate the probability of a batter getting 0, 1, 2, 3, 4, or 5 hits in 5 at-bats.

```
dist = stats.binom(5, 0.3) ①
df = pd.DataFrame({
    "hits": range(6),
    "probability": [dist.pmf(hits) for hits in range(6)], ②
})
df
```

① The first argument is the number of trials n and the second argument is the probability of success p.
② The *pmf* function returns the probability mass function for the given number of hits. This is the only difference in the object methods compared to continuous distributions. The analogous function for continuous distributions is *pdf*.

Output

```
   hits  probability
0     0      0.16807
1     1      0.36015
2     2      0.30870
3     3      0.13230
4     4      0.02835
5     5      0.00243
```

Figure 5.9 shows the graphs of the functions that describe the binomial distribution.

We can see from the table that the probability that our batter gets three hits in five at-bats is 0.1323 (`dist.pmf(3)`). If we want to know the probability that he gets three or fewer hits, we can use the cumulative distribution function (*cdf*).

```
print(f"probability of 3 hits: {dist.pmf(3):.4f}")
print(f"probability of 3 or fewer hits: {dist.cdf(3):.4f}")
print(f"probability of 3 or more hits: {dist.sf(2):.4f}") ①
```

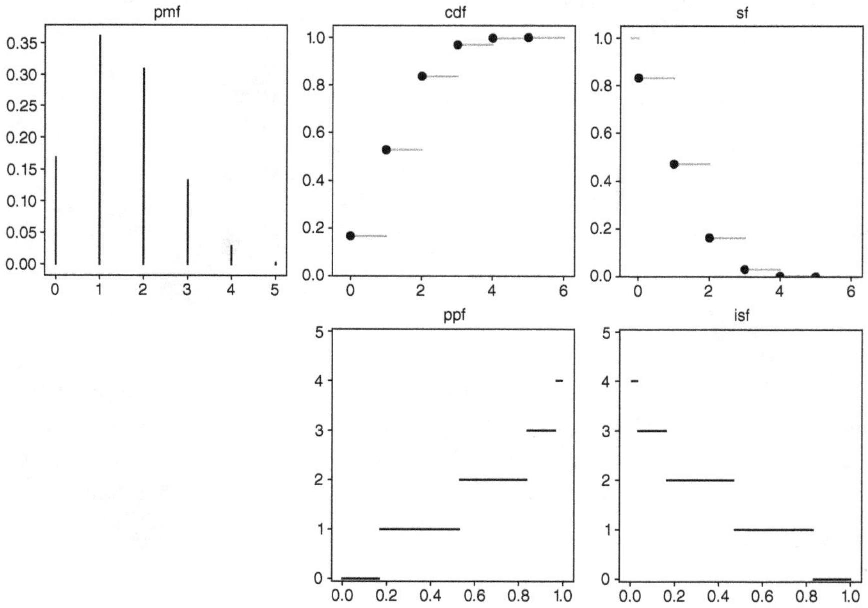

Figure 5.9 Probability mass function (pmf), cumulative distribution function (cdf), survival function (sf = 1-cdf), percent point function (ppf), and inverse survival function (isf) for the binomial distribution with $n = 5$ and $p = 0.3$.

① Note that we pass 2 to the *sf* function. The survival function tells us the probability of seeing three or more after we've already seen two.

Output
```
probability of 3 hits: 0.1323
probability of 3 or fewer hits: 0.9692
probability of 3 or more hits: 0.1631
```

We can compare these results to a resampling experiment.

```
random.seed(321)
hits = [random.choices([0, 1], weights=[0.7, 0.3], k=5).count(1) ①
        for _ in range(1000)]
hits = np.array(hits) ②
print(f"probability of 3 hits: {sum(hits == 3) / len(hits):.4f}")
print(f"probability of 3 or fewer hits: {sum(hits <= 3) / len(hits):.4f}")
print(f"probability of 3 or more hits: {sum(hits >= 3) / len(hits):.4f}")
```

① There is a lot going on in this one line; let's break it down. The *choices* function returns a list of 0s and 1s (*hit*). The `weights` argument defines the probability of each outcome. The `k` argument defines the number of draws (5). The result is a list of five 0s and 1s. We count the number of 1s using the *count(1)* function. We repeat this 1000 times and store the results in a list. Instead of using the *count* function, we can also use the *sum* function on the list of five values.
② The conversion to a `numpy` array is not strictly necessary, but it makes it easier to work with the data. For example, we can use the == or other comparison operators to compare the array to a value. This returns an array of `True` and `False` values. We can use the *sum* function to count the number of `True` values.

Output
```
probability of 3 hits: 0.1430
probability of 3 or fewer hits: 0.9760
probability of 3 or more hits: 0.1670
```

The resampling probabilities are very close to the theoretical results.

Exercises

5.1 In basketball, some fouls result in "free throws" (unimpeded shots from 15 feet away from the basket) by the player fouled. Over his career, a certain basketball player has scored on 1210 free throw attempts and missed 214 free throw attempts. What is his estimated probability of scoring on a free throw attempt?

5.2 Consider the following data on the median home value (in $000) on Boston neighborhoods (from the mid-20th century): (22, 13.1, 17.8, 20.3, 15.4, 11.7, 25.3, 15.2, 27.1, 23.2, 23.1, 18.1, 32.9, 20.3, 21.1, 21.1, 19.9, 23.1, 16.1, 10.4). Find the standard Normal score for the first value (22). You can consider this to be a sample from a larger population of neighborhoods.

5.3 Dice are small cubes held in the hand and then dropped or thrown on a flat surface, as part of a game. Each surface of a die has a number of dots, ranging from one dot to six dots. The number of dots shown on the top surface after a die lands is termed "how the die lands." If you throw a single die once,
a) What is the probability that it will land 3?
b) What is the probability that it will land 1 *or* 6?
c) What is the probability that it will land 1 *and* 6?

5.4 Geologists can predict how much oil a well produces, but there is uncertainty in the prediction. So, they express their prediction in probabilistic terms (volume in barrels per day):

Probability	**Volume**
0.40	75
0.45	90
0.15	125

What is the expected value of the well's production (in barrels per day)?

5.5 If you look at the scores on a collection of typical IQ tests, they will have a mean near 100 and a standard deviation near 15.
a) DISCUSSION: Why do you think this is so?
  i) This is the natural state of human intelligence
  ii) The tests are engineered to make it so
b) DISCUSSION: Is "IQ" the same thing as "IQ test score?"
c) Convert an IQ score of 130 to a *z*-score
d) If you were to select a person at random, what is the probability that they would have an IQ score of 130 or more, assuming the scores are distributed Normally?

6

Categorical Variables

In the previous chapter, we looked at binomial (two-outcome) variables. In this chapter, we expand that discussion to multi-category variables and relationships between categorical variables. After completing this chapter, you should be able to

- summarize categorical data in two-way tables
- calculate conditional probabilities
- perform Bayesian calculations
- perform tests of independence
- use the multiplication rule
- explain Simpson's paradox

6.1 Two-way Tables

We start with the data on UC Berkeley graduate admissions that were introduced in Chapter 3, looking first at a breakdown by gender.

Table 6.1 is a "2-way" table—it portrays subjects by their status on two variables—gender and admission status. It shows that the admission rate for men is a lot higher than the admission rate for women. More generally, tables like this are known as R × C tables (for row by column) or contingency tables (because you can read counts for one variable contingent on the other variable taking a certain value).

In Table 6.2, we see these data as a percentage table, which makes clearer the difference between women and men with respect to admission rates.

Statistics for Data Science and Analytics, First Edition. Peter C. Bruce, Peter Gedeck, and Janet Dobbins.

Companion website: www.wiley.com/go/Wiley_Statistics_for_Data

Table 6.1 Applications to UC Berkeley departments.

	Female	Male	All
Admitted	557	1198	1755
Rejected	1278	1493	2771
All	1835	2691	4526

Table 6.2 Applications to UC Berkeley departments.

	Female	Male	All
Admitted	30.35	44.52	38.78
Rejected	69.65	55.48	61.22
All	100.00	100.00	100.00

6.2 Conditional Probability

Looking at one variable at a time leads us to the concept of "conditional probability"—calculating a probability for one variable, conditional (contingent) on the value of another. Let's explore that further by looking at the UC Berkeley data by department.

Table 6.3 shows the relationship between the gender of the applicant and the department to which the person applied.

The gender variable is in the rows and has two categories, while the department variable is in the columns and has six categories. We call this a 2×6 contingency table because it has two rows and six columns and shows all the possible combinations of gender and department.

The cells in this table contain counts, so 593 females applied to Dept. C. The cells must be mutually exclusive and jointly exhaustive: every application has to

Table 6.3 Numbers of applicants to UC Berkeley departments A, B, C....

	A	B	C	D	E	F	All
Female	108	25	593	375	393	341	1835
Male	825	560	325	417	191	373	2691
All	933	585	918	792	584	714	4526

go in one and only one cell. One person could apply to more than one department, in which case there would be an entry in the table for each application.

Here are some probabilities we can read from Table 6.3. The probability that a randomly selected application is from a female is 1835/4526. The probability that the application is to Dept. E is 584/4526. The probability that the application satisfies both descriptions—female and Dept. E—is 393/4526.

Try It Yourself

From the above table, find the probability that

1) An application is from a female and applies to Dept. A.
2) An application is from a female.
3) An application is to either Dept. A or Dept. B.
4) An application to Dept. B is from a female.
5) An application from a female is to Dept. B.

Questions four and five above are both conditional probability questions—they ask for the probability of one event, given another event. For example, given that an application is from a female (event one), what is the probability that the application is to department B (event two)?

There were 1835 applications from females, and 25 of them were to department B. This is 1.36% or a probability of 0.0136. This probability is "conditioned" on the knowledge that the applicant is female.

Given that an application is to department B, what is the probability that the applicant is female?

There were 585 applications to department B, and 25 of them were from females. This is 4.27%, or a probability of 0.0427. This probability is conditioned on the knowledge that the person has applied to department B.

We implicitly calculate conditional probabilities all the time. Your favorite soccer team has won 57% of its games, but your estimate of their probability of winning their next game will be lower if you know that their star striker is injured and can't play.

The probability that a streaming service customer rents a new movie might be 0.0001. That probability rises a lot if we know that the person watches a trailer for the movie—the probability of a purchase, given that someone has watched the trailer, is higher than the unconditional probability of a purchase.

The notation for this relationship of the probability of A, given B, is a vertical line.

$$P(\text{rent}|\text{trailer}) = \text{probability of rental, given watched trailer}$$

In arithmetic terms, what happened is that the denominator in the conditional probability became a lot smaller—it includes only the people who watched the trailer. The formula for a conditional probability is

$$P(A|B) = \frac{P(A \cap B)}{P(B)}$$

In terms of the trailer/rent example, the probability of a rental (A), given trailer (B), is the probability of the rental AND the trailer, divided by the probability of watching the trailer. Put in terms of numbers, it is the number of people who watched the trailer AND rented divided by the number of people who watched the trailer. Put in terms of percentages, it is the percentage of trailer-watchers who rent.

6.2.1 From Numbers to Percentages to Conditional Probabilities

We can explicitly present the conditional probabilities in two additional tables. Table 6.4 breaks down each department's applications by gender in percentages.

Try It Yourself
Describe what you see in the table. Do the departments all have about the same ratio of males to females, or do they differ quite a bit? If they differ, which departments differ most from what you might expect? Be sure to account for a mostly male overall applicant pool, which means you would not expect the department ratios to be 50/50.

From Table 6.4, we can say that the probability is 88.42% that an application to Department A is from a male. If you go back to the original tables of counts, you can see that this was computed as 825/933. The notation here would be $P(M|A)$—the probability of M given A, or in more explicit English, the probability of picking a male when choosing among the applications to Department A.

Table 6.4 Percentage of department applications by gender.

	A	B	C	D	E	F	All
Female	11.58	4.27	64.6	47.35	67.29	47.76	40.54
Male	88.42	95.73	35.4	52.65	32.71	52.24	62.08
All	100	100	100	100	100	100	100

Table 6.5 Male/female applications by department.

	A	B	C	D	E	F	All
Female	5.89	1.36	32.32	20.44	21.42	18.58	100
Male	30.66	20.81	12.08	15.50	7.10	13.86	100
All	20.61	12.93	20.28	17.50	12.90	15.78	100

Try It Yourself
From Table 6.4, find $P(M|C)$, $P(\sim M|D)$ and $P(M)$.

The second conditional probability, Table 6.5, breaks down each gender's applications by department. From this, we can see that $P(A|M) = 37.88\%$.

Note that the two conditional probability tables above break down the applications in different ways, and the values found in one table are nowhere to be found in the other one.

In plain English, Table 6.4 shows that almost all the applicants to Department A are male, but Table 6.5 shows that only about a third of the males apply to Department A.

Getting these two probabilities mixed up is a common error in using and interpreting probabilities.

Try It Yourself
Find $P(E| \sim M)$ and $P(\sim M|E)$. Show where these numbers can be found in a table, and show what counts must be divided to get these numbers.

6.3 Bayesian Estimates

Consider a medical screening test that gives a positive result in 98% of the cases where the condition is present but also gives a false positive result in 3% of cases where the condition is not present. Suppose 0.1% of the people we screen actually have the condition (to pick an actual disease, this is the approximate prevalence of juvenile arthritis).

If you test positive, what is the probability that you have the disease?

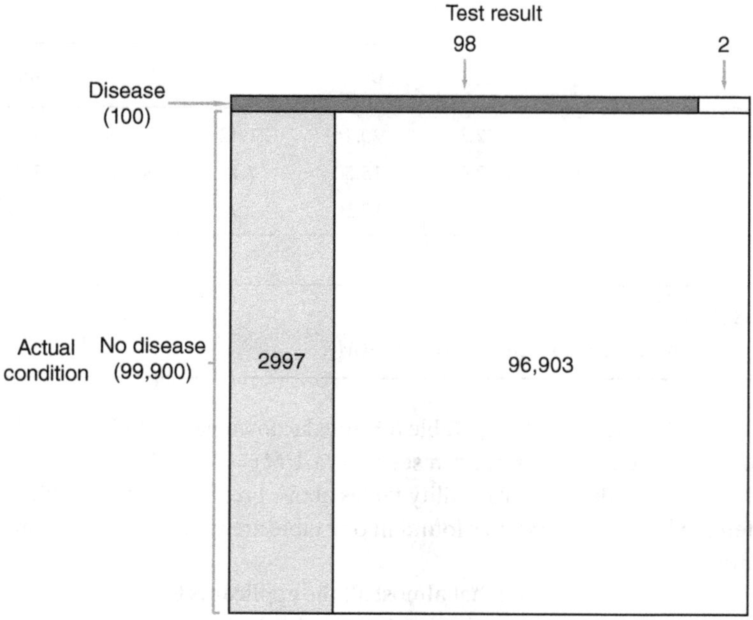

Figure 6.1 Bayesian calculation (medical test example).

To work out the problem, let's say you screen 100,000 people.

100 of them (0.1%) will actually have the condition, so the other 99,900 do not.

Of the 100 who have the disease, 98 test positive—giving 98 true positives. Two of those with the disease test negative, which is two false negatives.

Of the 99,900 who do not have the disease, 97% test negative, but 3% or 2997 show false positive results. So we have

- 98 true positives
- 2997 false positives
- 3095 total positives.

Therefore, even if you test positive, the probability is only 98/3095 = 0.0317 = 3.17% that you have the disease (Figure 6.1).[1]

6.3.1 Let's Review the Different Probabilities

Shifting now from percentages to probabilities:

1) 0.001: The overall prevalence of the disease is an unconditional probability.

1 A short video on these calculations is available on the book resources website.

2) 0.98: The probability of a positive test if you have the disease is a conditional probability—$P(\text{positive}|\text{disease})$.
3) 0.03: The probability of a positive test if you do not have the disease is a conditional probability—$P(\text{positive}|\text{no disease})$.
4) 0.0317: The probability of the disease if you have a positive test is a conditional probability—$P(\text{disease}|\text{positive})$.

Probabilities one through three are known to the researchers, but probability four is of primary interest. It is not known directly—we had to calculate it.

6.3.2 Bayesian Calculations

The calculations that we did to determine #4—the probability that you have the disease if you test positive—are termed Bayesian calculations.

There is a formula to calculate Bayesian estimates using probabilities 1–3 above, but it is quite formidable in appearance. The least confusing way to make and understand these estimates is to convert the percentages to actual numbers, as we did above when we said, "let's say you screen 100,000 people."

The essence of Bayesian estimation is that you have some initial or prior estimate of a probability—the known overall prevalence of the disease. You then receive some pertinent information—the test result—and revise the initial estimate.

In this case, the revision of the initial estimate is surprisingly small and potentially confusing. Faced with a test that is 98% effective in identifying the diseased cases, most people—and many doctors!—have a hard time believing that if you test positive, you have only a 3% chance of having the disease.

This confusion represents a real problem for mass screenings in populations; the more so, the rarer the condition. While you do get true positives for most of the folks having the condition, there are so many more folks who do not have it that their false positives swamp the real positives. This is the reason you often find statisticians testifying against mass screening proposals. To put a human face on it, imagine that folks who test positive will lose their job, be denied insurance, be barred from professional sports, or be told they have AIDS when that is the right decision in only a small percentage of cases.

Public health authorities go back and forth on the right level of screening for common diseases, as more is learned about offsetting costs and benefits, and as the public weighs in.

6.4 Independence

We often see the question posed of whether two variables are associated—whether smoking is correlated with cancer, obesity with diabetes, or home runs with strikeouts (in baseball). The machine learning revolution is predicated on the notion that we can predict unknown variables by their association with known variables. The opposite notion—lack of association—is called independence. In this section we compare the graduate school departments by admission rates (Table 6.6).

Try It Yourself
What is the probability of being admitted if someone applies to Dept. A? What is the admission probability if someone applies to Dept. F?

Is the admission rate independent of the choice of department?

We would say that admission to a department is independent of the choice of the department if the admission rate is the same no matter which department you choose. In symbols:

$$P(\text{Admit}|\text{Department}) = P(\text{Admit})$$

Definition: Independent Events Two events are independent of one another if the probability that one will occur is unaffected by whether or not the other one occurs.

Whether New York beats Chicago in a baseball game is probably unaffected by whether Los Angeles beats San Diego in a different baseball game—the two events are independent.

On the other hand, whether it rains on Tuesday in Paris is probably associated with whether it rains in Paris on Monday—the two events are connected and not independent.

Table 6.6 Admission rates (percent) by department.

	A	B	C	D	E	F	All
Admitted	64.42	63.25	35.08	33.96	25.17	6.44	38.78
Rejected	35.58	36.75	64.92	66.04	74.83	93.56	61.22
All	100	100	100	100	100	100	100

⚠ Analysts often assume independence when it is not warranted, because, without that assumption, analysis becomes difficult or impossible. But no analysis may be better than a bad analysis.

The financial collapse of 2008 was prompted, in part, by poorly designed financial products that were based on bundles of subprime mortgages. The default risk of these bundles was confidently predicted to be at a comfortably low level, based partly on the assumption that the failure of mortgage A did not affect the probability that mortgage B would fail. In fact, this was not strictly true—systemic problems, including fraud and gross negligence, affected most subprime mortgages, and their outcomes were quite correlated.

What about the admission rates? The variables Admit and Dept. would be independent if the probability of admission were the same for every department. Clearly, they are far from independent here—the probability of admission varies substantially from department to department.

Try It Yourself

This table shows counts for 100 possible outcomes of events A and B.

	A	$\sim A$
B	20	20
$\sim B$	40	20

Are A and B independent?

6.4.1 Chi-square Test

Just as with the hospital error reporting, we would like to know whether an apparent association between categorical variables is real or possibly the result of chance. The statistical test we use for this is a test for independence. Let's illustrate with the sensor components for a self-driving car.

6.4.1.1 Sensor Calibration

Self-driving cars are guided by a variety of cameras and other sensors that provide data to the navigation algorithm. Over time, due to bumps, vibrations, and use, the sensors can lose their accuracy and must be recalibrated. As part of the ongoing research into recalibration, an automaker has two teams, one in California and one in Texas, perform an identical battery of tests to determine whether a vehicle's sensors need recalibration. Recalibration is done on a per-vehicle basis—the sensors must work together, so if any require recalibration, all must be recalibrated. Here are the results (whether a vehicle's sensors must be recalibrated) for 200 vehicles (same model) after 50 hours of driving time:

Recalibrate?		
Texas	25 yes	175 no
California	17 yes	183 no

Is the need for recalibration independent of state to a statistically significant degree? If so, then the automaker might plan recalibrations without worrying about the test center location. If, on the other hand, recalibration rates differ significantly between Texas and California, then the company will have to worry about either differences in driving conditions or differences in test center procedures in setting recalibration policies.

For a more specific formulation of this test, we start by asking, "what distribution of recalibrations would we *expect* to see if the two states share a common recalibration rate, and might the *actual* distribution differ just by chance?" If the state testing centers shared a common rate (i.e. the null hypothesis), we would expect the total 42 recalibrations to be split equally.

Expected results with a common rate:

Recalibrate?		
Texas	21 yes	179 no
California	21 yes	179 no

We see that Texas had four more recalibrations than expected and four fewer non-recalibrations, and California is just the reverse. From this, we can construct a table of the extent to which the actual results differ from expectations:

Source: Sundry Photography/Adobe stock

Recalibrate?		
Texas	4 yes	−4 no
California	−4 yes	4 no

For an overall measure of "departure from expected," we sum up the absolute differences for a total of 16. We ask for the *absolute difference* because we are interested in an overall departure from expectation, not whether Texas, in particular, yields more recalibrations. Is 16 more than we might expect by chance? To test, we can:

1) Put 42 1's (yes) and 358 0's (no) in a hat to represent the hypothesized common recalibration rate.
2) Shuffle the hat and deal the numbers out in two resamples of 200 each.
3) Count the number of 1's and 0's in each resample.
4) Sum up the absolute differences between this resampled result and the expected result under the assumption of a common rate.
5) Repeat 2–4 many times.
6) How often did we get a difference as great as 16?

If we often get a resampled sum as great as 16, we can't rule out chance as an explanation for the difference between the two states.[2]

You may have noticed that it is not strictly necessary to consider all four cells in the 2×2 table; actually, one will suffice. This is not so with tables that have more than two rows or columns, which we take up in Chapter 8.

6.4.1.2 Standardizing Departure from Expected

In 1900, Karl Pearson, one of the founding fathers of statistics, introduced a formula-based approximation based on a standardized "departure from expectation" test statistic—the chi-square test. The process is analogous to the normalization of data, where we subtracted the mean and divided by the standard deviation (see Section 5.4.1). This "normalized" departure from expectation statistic differs from the "raw" departure from expectation statistic that we calculated above:

1) Squared differences, rather than absolute differences, are used (in either case, both positive and negative differences are rendered positive).
2) Each difference is divided by the expected value for that cell.

---

2 In our procedure, we could have replaced each number in the hat before dealing out the next number; this is called resampling with replacement. Both procedures have been used in multi-sample permutation tests like this; they have slightly different statistical properties that are beyond the scope of this text.

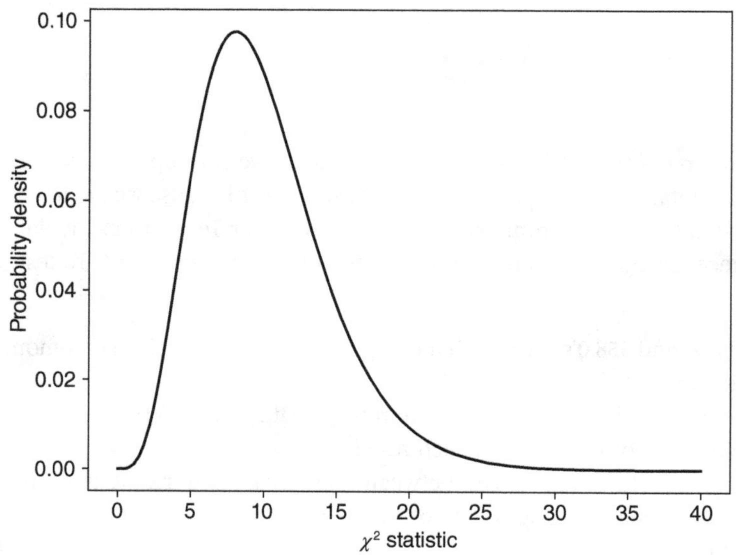

Figure 6.2 Example χ^2 distribution for 10 degrees of freedom.

The formula for the chi-square statistic is:

$$\chi^2 = \sum_{i=1}^{n} \frac{(O_i - E_i)^2}{E_i}$$

This *chi-square* statistic can then be compared to a standard chi-square distribution, as generated by repeated random shuffling of a null model. In precomputer days this was important, as this standard distribution only needed to be calculated once.

The distribution is long-tailed to the right, and its general form is shown in Figure 6.2. The exact shape of the distribution differs, depending on the degrees of freedom, which are governed by the number of rows and columns in the table. The example is for 10 degrees of freedom.

6.5 Multiplication Rule

Now that we have a test for independence, let's turn to a simple rule for calculating the probability that both events *A and B* will occur. In general, the probability of A and B occurring together is the probability of (A) times the probability of (B, given A).

Multiplication Rule—General Form:

$$P(A \cap B) = P(A) \times P(B|A)$$

If A and B are independent of one another, knowing that A has happened makes no difference in calculating the probability of B. In this case, $P(B) = P(B|A)$. So, when events A and B are independent of one another, the multiplication rule simplifies to this:

Multiplication Rule—Independent Events:

$$P(A \cap B) = P(A) \times P(B)$$

The probability of A and B is the probability of A times the probability of B.

For example, if we take the two baseball games referred to earlier, whose outcomes are independent, let's say we have the following probabilities:

- P(New York beats Chicago) = 0.60
- P(Los Angeles beats San Diego) = 0.45

Then the probability that New York wins AND Los Angeles wins = 0.60 × 0.45 = 0.27.

The multiplication rule is prone to misuse. The misuse is to blindly assume two events are independent and then use this formula to compute the probability that both happen. For example, in Los Angeles in 1964, a robbery case was tried in which eyewitness testimony described the assailants as a bearded Black man traveling in a car with a blond woman. Prosecutors contended that the probability of seeing a woman with blond hair was 1/3, a Black man with a beard was 1/10, and an interracial couple in a car was 1/1000. Adding additional eyewitness information and using the multiplication rule, they estimated the probability of some other random couple meeting these descriptions as 1/12,000,000; therefore, it was most likely the defendants. There were numerous problems with the analysis, but a big one was that the events cited are not independent: once you have a blonde woman and a Black man, the probability of an interracial couple is no longer 1/1000—it is 1.0.[a]

For events that are *not* independent, use the more general form of the multiplication rule above.

[a]Cited in *Statistics*, 2nd ed., Freedman, Pisani, Purves and Adhikari, 1991, W. W. Norton, p. 219.

The multiplication rule holds only when A and B are independent. It works the other way as well: if this equation is true, then the events are independent.

6.6 Simpson's Paradox

The UC Berkeley data are the most famous example of Simpson's Paradox. It is named for Edward Simpson, a British statistician who got his start in statistics at Bletchley Park—the World War II British decoding center (Figure 6.3).

Table 6.7 shows the admission rates for men and women at UC Berkeley graduate schools.

The gender and the admission of an applicant do not seem to be independent for the college as a whole. Women have a lower rate of admission than men. Is this evidence of discrimination against women?

When you look at individual departments (Table 6.8), though, you do not see the apparent discrimination visible in the overall figures in Table 6.7.

Women had higher admission rates in every department except C and E, where their disadvantage was slight.

What's going on? How can women have a higher rate of admission in nearly every department but a lower rate overall? Let's look again at the actual number of applications for each department—originally presented at the beginning of Section 6.2 in Table 6.3.

Figure 6.3 Bletchley Park. Source: DeFacto/Wikimedia Commons/CC BY SA 4.0.

Table 6.7 Apparent discrimination against women at Berkeley.

	Female (%)	Male (%)	All (%)
Admitted	30.35%	44.52	38.78
Rejected	69.65	55.48	61.22

Table 6.8 Berkeley admission rates by department.

	A (%)	B (%)	C (%)	D (%)	E (%)	F (%)
Female	82.41	68.00	34.06	34.93	23.92	7.04
Male	62.06	63.04	36.92	33.09	27.75	5.90

Try It Yourself
Examine the percentages in Table 6.8. Then, examine Table 6.3 and where men and women tend to apply. Can you explain the paradox?

Simpson's Paradox is also termed the "aggregation paradox," apparent contradictons that emerge when putting parts together into a whole.

6.7 Python: Counting and Contingency Tables

6.7.1 Counting in Python

There are many ways of counting occurrences of items in a list. For example, we can count the number of heads in 100 coin flips by iterating over the list of coin flips.

```
# create a list of 100 coin flips
import random
random.seed(1234)
coin_flips = [random.choice(["H", "T"]) for i in range(100)]

count = 0 ①
for coin_flip in coin_flips: ②
    if coin_flip == "H": ③
        count += 1
print(count)
```

① Initialize a counter to zero.
② We iterate over all elements of our list of coin flips.
③ Every time we encounter a head (`"H"`), we increase the counter by one.

This can also be written as a list comprehension making use of the fact that `True` is equal to 1 and `False` is equal to 0.

```
sum(coin_flip == "H" for coin_flip in coin_flips)
```

Counting is such a frequent operation that Python even provides a built-in method for lists:

```
coin_flips.count("H")
```

The examples so far were only counting occurrences of a single item. We can also count occurrences of multiple items. For example, we can count the number of heads and tails in 100 coin flips. In this case it is convenient to use a dictionary to store the counts.

```
counts = {"H": 0, "T": 0} ①
for coin_flip in coin_flips:
    counts[coin_flip] += 1 ②
print(counts)  # prints: {'H': 55, 'T': 45}
```

① Initialize a dictionary with keys "H" and "T" and values 0.
② We iterate over all elements of our list of coin flips.
③ Using the coin flip outcome as the key, we select the corresponding counter and increase it by one. The += operator is a shorthand for `counts[coin_flip] = counts[coin_flip] + 1`.

If the number of items we want to count is large, we can use the *Counter* class from the `collections` module.

```
from collections import Counter
counts = Counter(coin_flips) ①
print(counts)  # prints: Counter({'H': 55, 'T': 45})
print(f'Number of heads: {counts["H"]}')  # prints: Number of heads: 55
```

① The *Counter* class takes an iterable as input and returns a dictionary with the counts of each item.

The resulting counts object can be updated with additional items using the *update* method. For example, we can count the number of heads and tails in 100 coin flips and then update the counts with 100 additional coin flips.

```
counts = Counter(coin_flips)
print(counts)  # prints: Counter({'H': 55, 'T': 45})
counts.update(random.choice(["H", "T"]) for i in range(100))
print(counts)  # prints: Counter({'H': 101, 'T': 99})
```

6.7.2 Counting in `Pandas`

The `pandas` package provides the *value_counts* method to count occurrences of values in a column of a *pd.DataFrame* or a *pd.Series*.

```
import pandas as pd
df = pd.read_csv("microUCBAdmissions.csv")
counts = df["Admission"].value_counts() ①
print(f"Number of admitted students: {counts['Admitted']}") ②
counts
```

① The *value_counts* method returns a *pd.Series* object with the counts of each value. The values are sorted from most frequent to least. The index of the *pd.Series* object contains the values.
② We can access the counts for a specific value by using the value as the key.

Output
```
Number of admitted students: 1755

Admission
Rejected    2771
Admitted    1755
Name: count, dtype: int64
```

The *value_counts* method can also be used with multiple columns. In this case, the resulting *pd.Series* object is a multi-indexed series.

```
counts = df[["Admission", "Gender"]].value_counts() ①
counts
```

① We provide a list of column names to create a subtable that contains only the columns we are interested in.

Output
```
Admission  Gender
Rejected   Male      1493
           Female    1278
Admitted   Male      1198
           Female     557
Name: count, dtype: int64
```

There are various ways to access the counts in a multi-indexed `pandas` object. Here are a few examples:

```
print(f"Number of admitted male students: {counts['Admitted', 'Male']}") ①
print(counts["Admitted"]) ②
print(counts[:, "Female"]) ③
```

① Using a tuple of keys returns a specific value. The keys need to be in the order of the columns, here Admission, Gender.
② If we only provide a few keys, the remaining keys are assumed to be all included. In our case, we specify Admission, but show all values associated with Gender. This is equivalent to `counts["Admitted", :]`.
③ Using `:` for a key, will include all the values associated with that key.

Output
```
Number of admitted male students: 1198

Gender
Male      1198
Female     557
dtype: int64

Admission
Rejected    1278
```

```
Admitted     557
dtype: int64
```

The *value_counts* method can also be used to calculate relative frequencies by setting the argument `normalize=True`.

```
counts = df[["Admission", "Gender"]].value_counts(normalize=True)
counts
```

Output

```
Admission  Gender
Rejected   Male      0.329872
           Female    0.282369
Admitted   Male      0.264693
           Female    0.123067
dtype: float64
```

Sometimes, it can be useful to convert the resulting *pd.Series* object to a *pd.DataFrame* object. This can be done with the *reset_index* method. It will convert the index into one or more columns.

```
counts.reset_index()
```

Output

```
  Admission  Gender         0
0  Rejected    Male  0.329872
1  Rejected  Female  0.282369
2  Admitted    Male  0.264693
3  Admitted  Female  0.123067
```

6.7.3 Two-way Tables Using `Pandas`

To generate a two-way table, we will need to count occurrences of pairs of items. We can use the approach from the previous section and store the counts in a dictionary. However, this quickly becomes tedious. Instead, we can use the *pd.crosstab* method from the `pandas` package. Let's see how we can recreate the first two-way table from Section 6.1 using *pd.crosstab*.

```
df = pd.read_csv("microUCBAdmissions.csv")
pd.crosstab(df["Admission"], df["Gender"])
```

The *pd.crosstab* method takes two (or more) arguments. With two arguments, the first is the variable to be used for the rows, and the second is the variable to be used for the columns. The resulting *pd.DataFrame* object contains the counts of each combination of the two variables. The result is:

```
Gender        F     M
Admission
```

```
Accepted     557  1198
Rejected    1278  1493
```

It is not quite what we've seen in Section 6.1. The sums of the columns, rows, and the full table are missing. We can add them by using the `margins=True` argument.

```
pd.crosstab(df["Admission"], df["Gender"], margins=True)
```

Now we get the same information.

```
Gender          F      M   All
Admission
Accepted      557   1198  1755
Rejected     1278   1493  2771
All          1835   2691  4526
```

The *pd.crosstab* method can also be used with more than two variables; see the `pandas` documentation for example.

In Chapter 6, we also came across two-way tables that expressed conditional probabilities. The function *pd.crosstab* can return conditional probabilities with the `normalize` argument. The normalize argument can take the values `"all"`, `"index"`, or `"columns"`. Using `"all"` normalizes the full table.

```
pd.crosstab(df["Admission"], df["Gender"], normalize="all",
margins=True)
```

Each value in the two-way table is divided by the total number of observations. The result is:

```
Gender         Female       Male       All
Admission
Admitted     0.123067   0.264693   0.38776
Rejected     0.282369   0.329872   0.61224
All          0.405435   0.594565   1.00000
```

To get conditional probabilities, we need to set `normalize` to either `"index"` or `"columns"`. With `"index"`, the values in each row are divided by the sum of the row. This gives us the conditional probability of the column variable given the row variable.

```
pd.crosstab(df["Admission"], df["Gender"], normalize="index",
margins=True)
```

Output

```
Gender         Female       Male
Admission
Admitted     0.317379   0.682621
Rejected     0.461205   0.538795
All          0.405435   0.594565
```

Setting `normalize="columns"` returns the conditional probability of the row variable given the column variable.

```
pd.crosstab(df["Admission"], df["Gender"], margins=True,
normalize="columns")
```

Output

```
Gender        Female      Male       All
Admission
Admitted    0.303542  0.445188  0.38776
Rejected    0.696458  0.554812  0.61224
```

6.7.4 Chi-square Test

With the knowledge that we gained so far, we can perform the resampling experiment from Section 6.4.1 in Python. We start by creating a *pd.DataFrame* object with the observed counts and determine the common rate and the observed difference.

```
data = pd.DataFrame({
    "states": ["Texas"] * 200 + ["California"] * 200, ①
    "votes": ["yes"] * 25 + ["no"] * 175 + ["yes"] * 17 + ["no"] * 183
})
observed = pd.crosstab(data["states"], data["votes"])
common_rate = observed.sum(axis=0) / 2 ②
observed_difference = abs(observed - common_rate).sum().sum() ③
print(f"The observed difference is {observed_difference}")
```

① We use list operations here. `["Texas"] * 200` creates a list with 200 times the string `"Texas"`. The + operator concatenates the two lists.
② We use the *sum* method of the *pd.DataFrame* object to sum over the rows (`axis=0`). The result is a *pd.Series* object. We divide by 2 to get the common rate.
③ `observed - common_rate` subtracts a vector with two elements from a 2×2 matrix. How is this possible? The vector is repeated along the rows and columns of the matrix. This is called *broadcasting* and is a very useful feature. Be careful to check that the repetition is done in the way you expect. The double call of `sum()` first sums the columns to return a *pd.Series* object, and then sums the elements of the *pd.Series* object to return a single number. In future versions of `pandas`, you will only need to call `sum()` once.

Now, we can perform the resampling experiment.

```
import random
import numpy as np
random.seed(1234)
differences = []
votes = list(data["votes"]) ①
for _ in range(5_000):
    random.shuffle(votes) ②
    distribution = pd.crosstab(data["states"], votes) ③
    differences.append(abs(distribution - common_rate).sum().sum()) ④
at_least_observed = (sum(np.array(differences) >= observed_difference) /
                     len(differences))
```

```
print(f"Observed difference of at least {observed_difference}:"
      f"{at_least_observed:.1%}")
```

① We create a list with the votes from the original data, so that we can more easily reshuffle the votes.
② We use the *random.shuffle* method to create a randomized copy of the original votes.
③ Using the resampled votes, we create a new two-way table.
④ and calculate the absolute difference.

The outcome of our calculation tells us that in 25.9% of the resampled tables, the absolute difference is at least as large as the observed difference. This means that the observed difference is not unusual, so we cannot reject the null hypothesis that the voting behavior is independent of the state.

In Chapter 8, we will see that the chi-square test can be used to test for independence between two categorical variables. The `scipy` package provides a function to perform this test. The function takes a two-way table as input and returns the test statistic, the *p*-value, the degrees of freedom, and the expected counts.

```
from scipy import stats
result = stats.chi2_contingency(observed)
print(f"chi2 = {result.statistic:.3f}")
print(f"p-value = {result.pvalue:.4f}")
print(f"degrees of freedom = {result.dof}")
print("expected")
print(result.expected_freq)
```

The *p*-value of the test is 0.2536, which is very close to the result of our resampling experiment. The *chi2_contingency* function is also able to work with more complex multi-way tables. See the `scipy` documentation for more details.

Exercises

6.1 An energy company follows the practice of conducting a detailed survey of potential natural gas tracts before exercising lease options. The true state of a tract may be positive (economically recoverable natural gas is present) or negative (economically recoverable natural gas is not present). 35% of tracts are truly positive. The company's prior experience has been that a positive tract has a 70% chance of yielding a favorable geological survey, but a negative tract also has a 15% chance of yielding a favorable survey. If the company gets a favorable survey on a tract, what is the chance that it is a truly positive one?

6.2 Use the *pulse.csv* data for this question: Was the percentage of women who ran higher or lower than that of men?

6.3 Dice are small cubes held in the hand and then dropped or thrown on a flat surface as part of a game. Each surface of a die has a number of dots, ranging from one dot to six dots. The number of dots shown on the top surface after a die lands is termed "how the die lands."

a) If you throw two dice, is the probability of seeing a "1" on the first die independent of the probability of seeing a "1" on the second die?
b) If you throw two dice, what is the probability that you will see a "6" on the first die AND a "6" on the second die?
c) If you throw two dice, what is the probability that you will see a "6" on the first die OR a "6" on the second die?

6.4 In 2022, Earnshaw closed 56% of his sales opportunities (14/25), while Samuels closed only 54% (7/13). In 2023, Earnshaw again performed better—71% (10/14) compared with Samuels' 69% (20/29).

a) Calculate Earnshaw's and Samuels' performance over the entire two-year period. Who did better?
b) Explain the apparent contradiction between the individual year performances and the overall two-year performance.

6.5 A company offers a cloud service aimed at consumers, featuring a free level of service to all as well as higher service fee-based tiers. Currently, they offer a relatively low level of service for the free tier, hoping that people will eventually see the need for a higher fee-based level. However, if the service level is too minimal, people are likely to keep searching for something better. They decided to run an experiment, offering some customers more at the free level, hoping that more people would become engaged users and upgrade. For several weeks, new customers are randomly selected to receive either the "low" free service or the "high" free service. The number of upgrades over three months is tracked. Here are the results:

Plan	Initial Customers	Upgrades to "Pay"
Low level free	329	27
High level free	385	29

a) Calculate the difference between the two upgrade rates.
b) Specify a resampling procedure to test whether the difference in the upgrade rates might have arisen by chance.
c) Discuss: From a business perspective, what factors might enter into your decision about which way to proceed in the future?

6.6 Consider this hypothetical table on drug test results and employment status two years after the drug test for a sample of employees:

	Failed Drug Test	Passed Drug Test	Total
Still employed	7	89	96
Not employed	?	?	?
Total	16	?	173

a) Fill in the missing information in the table.
b) Is employment status independent of drug test results?

7

Surveys and Sampling

In this chapter, we discuss sampling as a device to gain a more accurate picture of your data. After you complete this chapter, you should be able to:

- Define statistical bias
- Explain the key feature of an ideal simple random sample (SRS)
- Use resampling to derive sampling distributions for proportions and means
- Calculate confidence intervals
- Define the bootstrap
- Describe sampling methods other than simple random sampling

7.1 Literary Digest—Sampling Trumps "All Data"

By the end of 1936, the United States was showing signs of economic recovery from the Great Depression that started with the 1929 collapse of Wall Street. GDP was back to where it was in 1929; it had fallen by a third in the interim. Unemployment was headed back to 15% after having risen to 25% at the depths of the recession. The good news was destined to be short-lived. Another recession hit in 1937, and the economy stagnated until World War II. However, in 1936, things were looking up for President Roosevelt as he campaigned for a second term. He had successfully put Social Security and Unemployment Compensation legislation through Congress and was banking on the popularity of his New Deal platform.

Political polling was in its infancy, though, so the preferences of the electorate remained somewhat obscure. In each presidential election year since 1916, *The Literary Digest*, a national weekly opinion magazine that often featured Norman Rockwell illustrations on its cover, had mailed out sample ballots to its readers and used the results to predict the outcome of the vote. *The Literary Digest* poll was a much-anticipated event; up through 1932, it had been accurate. In the summer of 1936, the *Digest* mailed out 10 million ballots and got back

Statistics for Data Science and Analytics, First Edition. Peter C. Bruce, Peter Gedeck, and Janet Dobbins.

Companion website: www.wiley.com/go/Wiley_Statistics_for_Data

2.4 million. On the strength of the results, the *Digest* predicted a landslide victory for Roosevelt's opponent—Republican Alf Landon.

Republican success in the September congressional elections in Maine seemed to validate *the Literary Digest*'s poll. Maine held some non-presidential elections early at the time and was regarded as a bellwether state, inspiring the catchphrase "As goes Maine, so goes the nation."

As it turned out, Roosevelt won in a landslide, capturing 62% of the popular vote and winning every state except Maine and Vermont. Maine's catchphrase was amended and became "As goes Maine, so goes Vermont." *The Literary Digest* went out of business shortly after the election.

What happened?

In addition to polling its readers, *The Literary Digest* also mailed ballots to lists of automobile owners and telephone subscribers. At a time when much of the nation was jobless and destitute, those who could afford magazine subscriptions, automobiles, and telephones were hardly representative of the population. They were wealthier and more Republican than the average voter, and they produced a biased prediction.

Definition: Bias A metric, estimate, or sampling procedure is statistically biased if, when you apply it to a sample from a population, it consistently produces over-estimates or under-estimates of a characteristic of that population.

In 1935, the year before *The Literary Digest* polled its readers, George Gallup, a young advertising executive with Young and Rubicam, founded the American Institute of Public Opinion. It was dedicated to the measurement of public opinion via statistically-designed surveys. He was convinced that what mattered was not the number of people surveyed but rather their representativeness—the degree to which they reflected the views of the general voting population.

The Gallup Poll

In October 1935, George Gallup published the first Gallup Poll—America Speaks. The first question was, "Do you think expenditures by the government for relief or recovery are too little, too great, or just about right?"

Sixty percent said, "too great."

Gallup quickly capitalized on the success of his political polling (see below). In the 1940s, he teamed up with David Ogilvy, famous for the phrase, "Advertising is neither entertainment nor an art form—it is a medium of information." Together, they worked with Hollywood executives to develop methods for predicting the box-office revenues of films based on measuring the appeal of the storyline, the popularity of the stars, the amount of publicity, and the reaction of preview audiences.

Figure 7.1 *The Literary Digest* was the premier literary and political commentary weekly of the early 20th century. Source: Agnes M. Watson/Wikimedia Commons/Public domain.

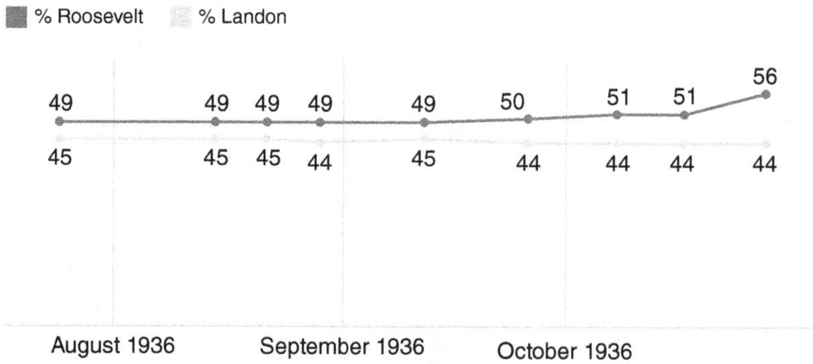

Figure 7.2 Gallup poll.

Gallup was convinced that 2000 people who were chosen scientifically would be a better predictor of electoral outcomes than millions chosen in the way that *The Literary Digest* did. He conducted bi-weekly polls, which showed Roosevelt leading by increasing amounts from August to October (Figure 7.2). The result can be seen on the Gallup organization's website, as the beginning of a continuum of similar tracking polls for US presidential elections up to the present.

Not only did Gallup correctly predict that Roosevelt would win, he also correctly predicted the outcome of *The Literary Digest* survey. He did this via a random sample that he selected to replicate, as far as possible, the demographics of the *Digest* survey respondents.

The key ingredient that catapulted Gallup to fame and success was the realization that a small representative sample is more accurate than a large sample that is not representative. There are now a variety of increasingly sophisticated devices that pollsters use to ensure representative results, but at the root of all of them lies random sampling.

7.2 Simple Random Samples

The most fundamental form of random sampling is the *simple random sample* (*SRS*). What is a simple random sample ?

The basic idea is that, in drawing such a sample, each element in the population has an equal chance of being selected. For a rigorous definition, though, we need more than just the idea of "random." For example, with a population of Democrats and Republicans, we could be lazy and say that we will flip a coin and, if it lands heads, our sample will consist of all the Democrats. Each member of the population has an equal—50/50—chance of being selected, but this procedure will hardly produce a representative sample. We need more.

Definition: Simple Random Sample (SRS) An SRS is produced by the equivalent of first placing the entire population, represented by slips of paper, in a box. Then, we shuffle the box and draw out the number of slips required for the sample. Statistically speaking, a sample of size n qualifies as an SRS if the sampling procedure affords each potential sample, i.e. each combination of n elements, an equal chance of emerging as the selected sample. The focus is on the procedure by which the sample is drawn, not on the characteristics of the resulting sample. It may, therefore, be more descriptive to use the term randomly drawn sample rather than a random sample.

Random sampling does not guarantee a completely representative sample. In fact, the use of random sampling almost guarantees that each sample will be a little different from the population from which it is drawn. The beauty of random sampling is that we can quantify the probable extent of this difference. We will see how in a moment, but for now, let's introduce or review some key concepts of the sampling process.

Definition: Population The **population** is the group that you are studying. It is often a somewhat amorphous concept that becomes difficult to define other than in broad terms. Consider the notion of New York voters. Does it include people who are eligible to vote but haven't registered? What about out-of-state students who attend university in New York and could vote there or at home?

Clearly, we need a working definition that we can put into practice.

Definition: Sampling Frame A **sampling frame** is a practical representation of the population—the slips of paper in the box from which we draw samples. For the New York voters, one possible sampling frame is the list of registered voters as of a given date.

Definition: Parameter A **parameter** is a measurable characteristic of the population—e.g. the mean, proportion, etc.

Definition: Sample A **sample** is a subset of the population. When it is randomly drawn, it is a **random sample**.

Definition: Random Sampling Technically, a random sampling process is one in which each element of a population has an equal chance of being drawn. You can think of it as a box with slips of paper that are well-shuffled, and you draw slips of paper blindly.

The words "statistic" and "statistics" have several meanings, all valid in different contexts. "Statistic," in the context of sampling, is defined as follows:

Definition: Statistic A **statistic** is a measurable characteristic of a sample, and it is used to estimate a population parameter.

Random Selection vs. Random Assignment: We began this book with a look at the effect that random variation in the *assignment* of hospitals to treatment or control might have had on the hospital no-fault error reporting study. In this chapter, we are examining the effect that random *selection* from a larger population might have on sample results.

7.3 Margin of Error: Sampling Distribution for a Proportion

You are probably familiar with the margin of error that often accompanies survey results, such as "42 percent think the country is headed in the right direction with a plus-or-minus two percent margin of error."

How is the margin of error calculated? What does it mean? The answer to the first question will help you understand the answer to the second.

The margin of error quantifies the extent to which a sample might misrepresent the population it is coming from, owing simply to the luck of the draw in who gets selected in the sample.

We'll start with this example.

In December 2010, a commercial polling organization sampled 200 US voters and found that only 72 voters, 36%, rated President Obama's handling of the economy positively—as good or excellent.

Definition: Point Estimate A point estimate is a statistic, such as a mean, median, etc., from a sample. The term "point" is to emphasize the fact that it is a single value, not a range of values.

Can we use a simulation to assess how reliable the 36% estimate from our sample is?

Before we answer that question explicitly, let's use a simulation to explore the extent to which the favorable proportion might change from resample to resample. We can put our 72 positives (1's) and 128 negatives (0's) in a box and repeatedly draw resamples of size 200, seeing how the proportion of 1's changes from draw to draw.

Try It Yourself

Using Python, execute a computer equivalent of the following simulation.

1) Put 200 slips of paper in a box. Mark 72 as 1 and 128 as 0.
2) Shuffle the box, and draw out a number. Record the number, and put the number back.
3) Repeat step two 199 more times, and record the total number of ones.
4) Repeat steps two and three many times (say, 1000), recording the number of ones each time.
5) Produce a histogram of the results.
6) Without worrying about being too precise, fill in the blanks in the following statement: Most of the time, the proportion who rated the handling of the economy "positively" in the sample lies between __ and __.

What is the importance of using 200 slips of paper in step 1? Could you use, say, 36 1's and 64 2's? 18 1's and 32 0's?

Which step in the above simulation is essential in modeling the size of the original sample?

7.3.1 The Confidence Interval

Looking at the histogram in Figure 7.3, we can see that the resample results range from about 25.0% positive to about 46.0% positive. For now, we can ignore the outliers beyond this interval. We can quantify the uncertainty with an interval that includes the large majority—say 90% or 95%—of the resampling results. For example, we find a 90% interval—called a confidence interval—by locating the fifth and 95th percentiles of the resampling distribution. This interval encloses 90% of the resampling results. The 90% confidence interval is approximately from 0.30 to 0.41.

To represent the population, how do we know what to put in the box?

In our simulation above, we knew what was in the population, and we wanted to learn how the samples behaved. In reality, we know only the sample result—we do not know what the population holds.

What do we put in the box? We create our best guess simulated population, which is based on the observed sample—36% positive. If we wanted to have a box with all voters, that would be a box with 45 million positive slips and 80 million not-positive slips.

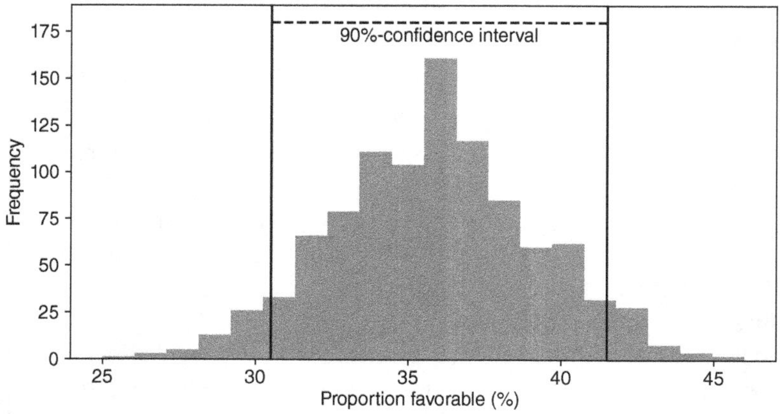

Figure 7.3 90% confidence interval; proportion "favorable" on x-axis.

7.3.2 A More Manageable Box: Sampling with Replacement

A box with millions of slips of paper is not manageable, and it is even a bit cumbersome on the computer. We use a shortcut instead:

1) A smaller hat—72 slips labeled "positive" and 128 labeled "not positive."
2) Sample *with replacement*.

Resampling *with replacement* ensures that the positive proportion in the box always remains the same. It is equivalent to resampling *without replacement* from a huge population. In the latter case, the positive proportion in the box remains pretty much the same from one draw to the next as long as the sample size remains very small relative to the population.

The container that holds the slips of paper is variously called a box, an urn, or a hat. The idea is the same.

7.3.3 Summing Up

To produce a confidence interval:

1) We can use the observed sample as a good proxy for the population.
2) The resample size should be the same as the original sample size.
3) The fact that the sampling is done with replacement allows the sample to serve, in effect, as a simulated population of infinitely large size.

7.4 Sampling Distribution for a Mean

We just calculated a confidence interval for a proportion. Let's do it now for the mean.

When you purchase a car, the dealer typically offers to purchase your old car, which is then resold. Toyota would like to know how much those used cars sell for—this is an important piece of information in revenue projection. Let's take a simple case—establishing the average resale price of the used cars disposed of by the dealers. The Toyota regional office in Europe takes a sample of recent Corolla sales, which yields the resale values shown in Table 7.1. The data are real sale values of used Toyotas; the scenario has been modified slightly.

Figure 7.4 is a histogram of the Toyota prices. The average sale price in this sample of 20 cars is 17,685 Euros. This is the *point estimate.* How much might this average be in error? In other words, if sales records could be located for all recent sales of used Toyota Corollas, how much might this estimate be off?

If we had easy access to records of all recent sales of used Corollas, we could just compare our sample to the entire data set. But we don't have easy access—that's why we took a sample.

Table 7.1 Toyota Corolla used car prices.

	13,500
	13,750
	13,950
	14,950
	13,750
	12,950
	16,900
	18,600
	21,500
	12,950
	20,950
	19,950
	19,600
	21,500
	22,500
	22,000
	22,750
	17,950
	16,750
	16,950
Mean:	17,685

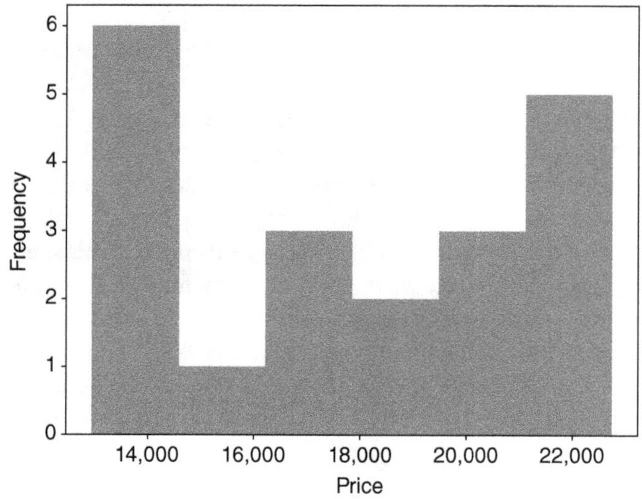

Figure 7.4 Histogram of Toyota prices.

One way to answer this question is to actually go out and sample additional sales transactions. Take another sample of 20, another sample of 20, etc., and see how much they differ from one another. But that will cost much more time, effort, and money than taking only a single sample of 20.

Can we take additional simulated samples instead of real samples like we did with the political poll? The trick, as always, is to determine what goes in the box. What population do we choose to sample from?

7.4.1 Simulating the Behavior of Samples from a Hypothetical Population

We can imagine creating a hypothetical population from our sample by replicating each item in our sample many, many times. We don't really know how big the total population of transactions is, but, as we saw earlier, it doesn't really matter as long as the sample is small relative to the population. We could then take resamples (without replacement) from this hypothetical population and see how those samples behave. We can achieve nearly the same thing by sampling *with replacement* from our original sample.

7.5 The Bootstrap

We have seen that individual sample results can vary, so a point estimate has uncertainty attached to it. Earlier, we discussed a political survey that asked 200 people their opinion of President Obama, and we used resampling to derive a confidence interval for the proportion favorable. We can do the same thing for measured data, like the Toyota prices: calculate a confidence interval for the mean.

Definition: A 90% Confidence Interval for the Mean A 90% confidence interval[1] is a range that encloses the central 90% of the resampled means using the simulation procedures described below. We then say that the population mean lies within the confidence interval with 90% confidence.

Let's review the Toyota Corolla case described above, where the average price in the sample—the sample mean—is 17,685 Euros.

7.5.1 Resampling Procedure (Bootstrap)

1) Write all 20 sample values on slips of paper, and place them in a box.
2) Draw a slip from the box, record its value, and replace the slip.
3) Repeat step two 19 more times, and record the mean of the 20 values, shown in Table 7.2.
4) Repeat steps two and three many more times, say 1000.
5) Arrange the 1000 resampled means in ascending order, and identify the fifth percentile and the 95th percentile—the values that enclose 90% of the resampled means. These are the endpoints of a 90% confidence interval, as shown in Table 7.3. Figure 7.5 is a histogram of the 1000 resampled means.

7.6 Rationale for the Bootstrap

We talked earlier about using the observed sample as a stand-in for the population by replicating it many times and placing the replicated slips of paper in a huge box. We noted that, in reality, we don't really need a huge box. Instead, we can achieve the same effect by sampling with replacement—putting each value back into the box after we have drawn it, thus yielding an infinite supply of each sample element. This is worth repeating here because the bootstrap engendered a lot of skepticism among statisticians when it was first developed, and many still puzzle over how we can get useful information by simply resampling from an observed sample.

Let's clarify some terms first.

1 We began this chapter by talking about surveys, where the term margin of error is frequently used. A margin of error is simply a plus or minus quantity attached to a point estimate, while a confidence interval is the actual interval. (If the confidence interval is not symmetric, it will differ from the margin of error.)

Table 7.2 Mean of 20 resampled or bootstrapped values.

	Original Sample	Resample
	13,500	13,750
	13,750	13,750
	13,950	21,500
	14,950	13,950
	13,750	16,750
	12,950	13,500
	16,900	20,950
	18,600	16,900
	21,500	17,950
	12,950	14,950
	20,950	16,900
	19,950	16,900
	19,600	13,750
	21,500	13,500
	22,500	21,500
	22,000	13,750
	22,750	19,950
	17,950	13,750
	16,750	16,900
	16,950	21,500
Mean	17,685	16,618

Table 7.3 90% confidence interval from percentiles of resampling distribution.

5%	Original Sample	95%
16,357	17,685	18,950

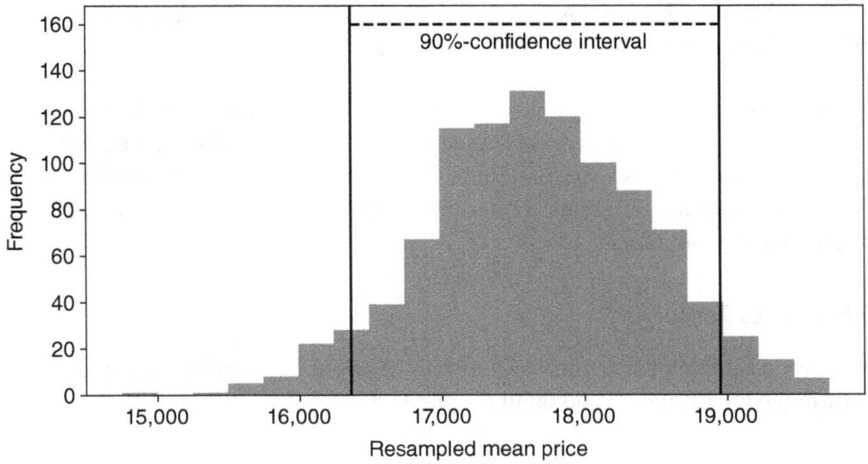

Figure 7.5 Histogram of used Toyota Corolla resale values.

Definition: Observation An observation is a data value for a single case. It could be a single value or multiple values, e.g. blood pressure and heart rate for a single subject.

Definition: Sample A sample is a collection of actual observations from a population.

Definition: Resample A resample is a new simulated sample, i.e. a collection of observations drawn from the original sample or generated randomly by a formula based on the original sample.

Definition: Sampling with Replacement When we sample with replacement, each item is replaced after it is drawn from a box, hat, etc.

Definition: Sampling Without Replacement In sampling without replacement, once an item is drawn, it is not eligible to be drawn again. Sampling without replacement is also called shuffling.

In the specific case described above, we have an original sample from a population, and instead of replicating the sample many times to create a hypothetical large population, we take resamples with replacement.

Definition: Single Simulation Trial A single simulation trial is the taking of a resample and performing further calculations with it. Typically, this means calculating the value of some statistic, such as the mean.

Definition: Simulation A simulation is a repeat of multiple single simulation trials and the collection of their calculation results.

Definition: The Bootstrap The *bootstrap* is a simulation in which we resample with replacement from an observed sample to observe the distribution of a sample statistic. This is a shortcut that eliminates the need to replicate the values in the sample many times and then sample without replacement from that large hypothetical population.

7.6.1 Let's Recap

We have a sample of size n from an unknown population, and we want to know how much an estimate based on that sample might be in error.

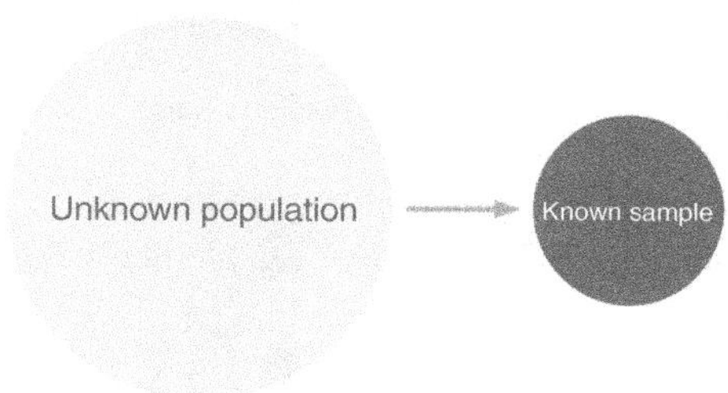

The key question is how do samples drawn from this population behave, i.e. how different are they from one another? We address this by taking resamples of size n, drawn with replacement from our sample (standing in for the population that the sample came from).

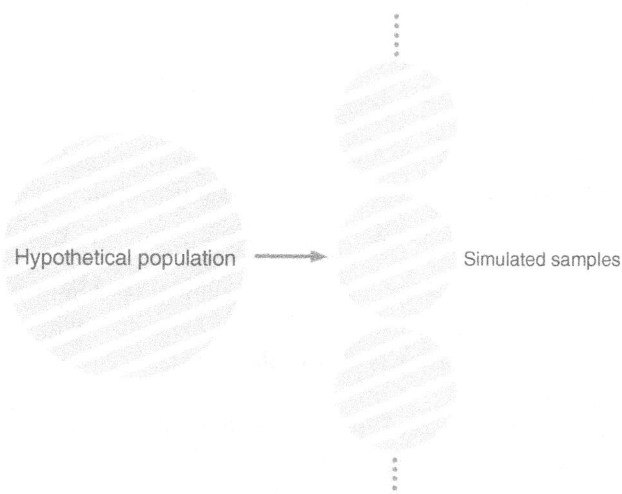

The accuracy of this procedure depends on how well the hypothetical population (in our imagination, the sample replicated over and over, but in reality, using the bootstrap)—mimics the characteristics of the unknown population.

Usually, the sample embodies all that we know about the population that it came from, and the bootstrap is an effective way to proceed. (It is also possible to create a simulated population from the sample parameters, such as a Normal population whose mean and standard deviation are estimated from the sample.)

Having found an appropriate representation of the population, we can draw resamples from that population, calculating the statistic of interest and recording it each time we draw a resample. Once again, the steps are as follows:

1) From our observed sample, calculate a statistic to measure some attributes of the population that we are examining.
2) Draw a resample from the original sample, with replacement, and record the statistic of interest.
3) Repeat step two many times.
4) Observe the sampling distribution of the statistic of interest to either learn how much our original estimate might be in error or how much it might differ from some benchmark value of interest.

7.6.2 Formula-based Counterparts to Resampling

Computational power and fast statistical software were not widely available until the 1980s, so resampling simulations were not feasible until then. Statisticians instead developed approximations that allowed analysts to calculate confidence intervals from formulas. These formulas rely on the fact that some sample statistics, like the mean, have a Normally-shaped distribution, even though the data the samples are drawn from are not Normally distributed.

The next sections cover scenarios for which we have already seen resampling solutions, and describe formulas that can also be applied in those situations. We provide them here as supplements because you are likely to see them in other contexts. Resampling and the formulas below are equivalent, alternate approaches. Formulas have the advantage that software provides ready-made routines to implement them. This can also be a disadvantage; they can be applied without a good understanding of whether and when they are appropriate. Formulas also have the disadvantage that data must meet certain conditions for the mathematical approximations to be suitable.

7.6.2.1 FORMULA: The Z-interval

To calculate a confidence interval for a mean, we revisit the standard Normal distribution we first saw in Chapter 4. Procedurally, the steps are as follows:

1) Find the values in a standard Normal distribution that correspond to the fifth and 95th percentiles. These values are −1.6449 and +1.6449.
2) Multiply by the sample standard deviation divided by the square root of the sample size. Then, add the sample mean.

Hint: If you Google Normal distribution probabilities, you will find web-based calculators for the above. With many such calculators, you can enter the sample standard deviation and mean and then obtain the interval directly.

The shaded area in Figure 7.6 represents the 90% z-interval for a Standard Normal distribution, i.e. a Normal distribution with mean = 0 and standard deviation = 1. We use this information to calculate a confidence interval.

The $100(1 - \alpha)\%$ z-interval for the mean =

$$\left(\overline{x} - z_{\alpha/2}\frac{s}{\sqrt{n}}, \overline{x} + z_{\alpha/2}\frac{s}{\sqrt{n}}\right)$$

where n is the sample size, $\overline{x}$ is the sample mean, s is the sample standard deviation, and $-z_{\alpha/2}$ and $z_{\alpha/2}$ are the z-values corresponding to the $\alpha/2$ percentile and the $1 - \alpha/2$ percentile.

7.6.2.2 Proportions

The same approach can be used for binomial (yes/no) data, where we use the proportion instead of the mean. Specifically, we use the Normal approximation to the

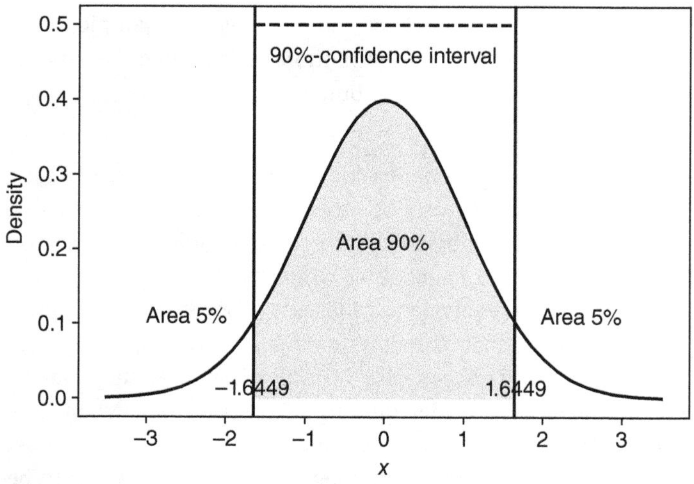

Figure 7.6 Normal distribution with mean = 0, SD = 1; x-axis is z-score.

binomial distribution that we introduced in the last chapter. The formula to calculate a 95% confidence interval around a given observed sample proportion p (where the sample is of size n):

$$p \pm 1.96\sqrt{\frac{p(1-p)}{n}}$$

In a Normal distribution, 95% of the data lie within 1.96 standard deviations of the mean, which, in this case, is the observed proportion. The second part of the equation,

$$\sqrt{\frac{p(1-p)}{n}},$$

is the standard deviation for a proportion. Had we wanted, say, a 90% confidence interval instead of a 95% interval, we would have multiplied by the coefficient that encloses 90% of the data: 1.645.

For the Normal approximation to the binomial to work effectively, the data should have at least five 1's and five 0's.

7.6.3 For a Mean: *T*-interval

The Normal distribution is actually not the best approximation to the true resampling distribution of the sample mean when samples have fewer than 30 values. The so-called *Student's t* distribution has the same general shape as the Normal and is almost indistinguishable for samples of size 30 or more. The t-distribution is actually a family of distributions whose shapes differ depending on the size of the sample. As sample sizes diminish, the t-distribution takes account of the greater variability with small samples and becomes lower and longer-tailed than the Normal. Since there are multiple t-distributions to be used when there are different sample sizes, a parameter called *degrees of freedom* must be specified when using the t-distribution. The number of degrees of freedom for the t-distribution is the sample size minus one.

The $100(1-\alpha)\%$ t-interval for the mean is

$$\left(\bar{x} - t_{n-1,\alpha/2}\frac{s}{\sqrt{n}}, \bar{x} + t_{n-1,\alpha/2}\frac{s}{\sqrt{n}}\right)$$

where n is the sample size, $\bar{x}$ is the sample mean, s is the sample standard deviation, $n-1$ are the degrees of freedom and $-t_{n-1,\alpha/2}$ and $t_{n-1,\alpha/2}$ are the t values corresponding to the $\alpha/2$ percentile and the $1-\alpha/2$ percentile.

7.6.4 Example—Manual Calculations

The t-interval calculations for the Toyota example are shown below.

The degrees of freedom for the one-sample t calculation are $n - 1 = 19$.

If $\alpha = 0.1$, then the fifth and the 95th percentiles for t_{19} are -1.7291 and 1.7291. These values traditionally were found in tables but now can be found in web-based calculators as well.

Using the same measures of $\bar{x} = 17{,}685$ and $s = 3507.252$ from the previous example, the 90% t-interval for the mean is

$$\left(\bar{x} - t_{19,0.05}\frac{s}{\sqrt{n}}, \bar{x} + t_{19,0.05}\frac{s}{\sqrt{n}}\right)$$
$$= \left(17685 - 1.7291\frac{3507.252}{\sqrt{20}}, 17685 + 1.7291\frac{3507.252}{\sqrt{20}}\right)$$
$$= (17685 - 1356.04, 17685 + 1356.04)$$
$$= (16328.96, 19041.04).$$

7.6.5 Example–Software

In Python, we can use the `scipy` package to calculate the 90% confidence interval using the *interval* method of the `stats.t` distribution.

```
import numpy as np
import pandas as pd
from scipy import stats

toyota = pd.read_csv("toyota.txt", header=None)
toyota.columns = ["price"]

# determine mean and its standard error
mean_price = np.mean(toyota["price"])
std_err_mean = stats.sem(toyota["price"])
dof = len(toyota["price"]) - 1
# calculate 90% confidence interval
ci_interval = stats.t.interval(0.9, dof, loc=mean_price,
   scale=std_err_mean)

print(f'Sample mean: {mean_price:.2f}')
print(f'Standard error of the mean: {std_err_mean:.2f}')
print(f'Degrees of freedom: {dof}')
print(f'90% confidence interval: [{ci_interval[0]:.2f}, '
      f'{ci_interval[1]:.2f}]')
```

Executing the code above produces the following output:

```
Sample mean: 17685.00
Standard error of the mean: 784.25
Degrees of freedom: 19
90% confidence interval: [16328.94, 19041.06]
```

The result is similar to but not exactly the same as the bootstrap interval.

7.6.6 A Bit of History—1906 at Guinness Brewery

The *t-distribution* was developed by William S. Gossett, a chemist at the Guinness brewery firm. In 1906, he took a leave of absence to study with the noted statistician Karl Pearson. When he returned to Guinness, one of his concerns was the reliability of conclusions drawn from small samples. In 1908, he published an article, "The Probable Error of a Mean," in the journal *Biometrika* (Figure 7.7).

Gossett published his article under the pseudonym Student because Guinness did not want competitors to know that they were employing statisticians. The existing industry practice at the time was to measure processes with very large samples, which was costly and time-consuming. Being able to work effectively with smaller samples would reduce cost and speed innovation, but this required knowing something about how reliable those small samples were.

Gossett started with a simulation (Figure 7.8). He worked with data on the physical attributes of criminals. Scientific society at the time was very interested in profiling and explaining the criminal "type," so lots of data were available from this captive audience. Gossett wanted to avoid data that would reveal that he worked for a brewery.

Gossett then took the recorded values for his samples, plotted them, and fitted curves to them. Figure 7.9 is his plot of the sample standard deviations, along with the curve that he fitted to them. (Notice that this is not Normally shaped; measures of variability tend to be distributed long-tail right.) The curve that he fitted to the resampled distribution of means is now known as the *t-distribution*.

If computing power like we have today had been available in 1908, it seems likely that computer sampling and simulation would have played a major role in statistics from the beginning. As it was, statisticians developed mathematical approximations as shortcuts to repeatedly dealing out cards.

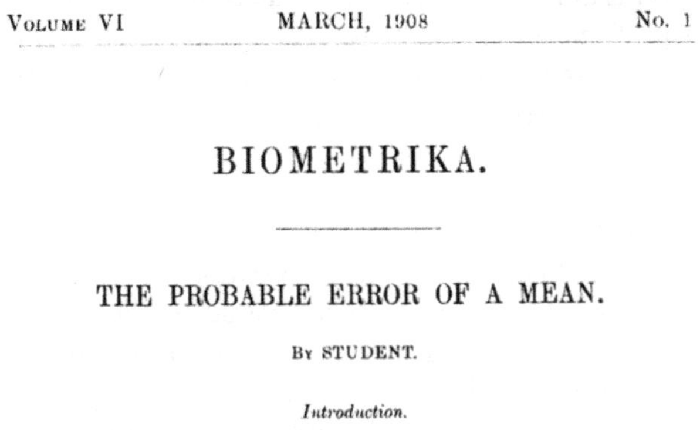

VOLUME VI MARCH, 1908 No. 1

BIOMETRIKA.

THE PROBABLE ERROR OF A MEAN.

BY STUDENT.

Introduction.

Figure 7.7 William S. Gossett's 1908 article.

SECTION VI. *Practical Test of the foregoing Equations.*

Before I had succeeded in solving my problem analytically, I had endeavoured to do so empirically. The material used was a correlation table containing the height and left middle finger measurements of 3000 criminals, from a paper by W. R. Macdonell (*Biometrika*, Vol. I. p. 219). The measurements were written out on 3000 pieces of cardboard, which were then very thoroughly shuffled and drawn at random. As each card was drawn its numbers were written down in a book which thus contains the measurements of 3000 criminals in a random order. Finally each consecutive set of 4 was taken as a sample—750 in all—and the mean, standard deviation, and correlation† of each sample determined. The difference between the mean of each sample and the mean of the population was then divided by the standard deviation of the sample, giving us the *z* of Section III.

Figure 7.8 Gossett"s description of his simulation.

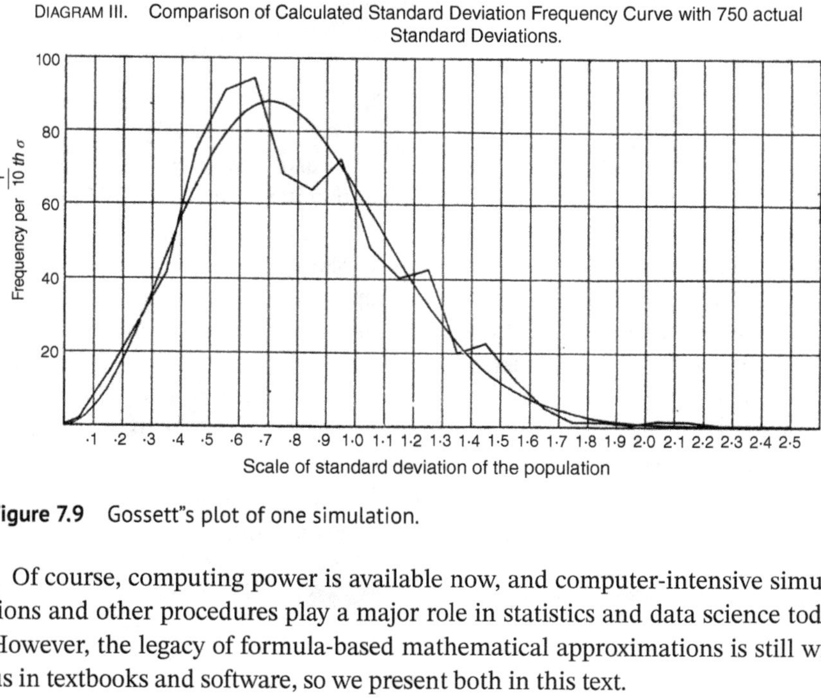

Figure 7.9 Gossett"s plot of one simulation.

Of course, computing power is available now, and computer-intensive simulations and other procedures play a major role in statistics and data science today. However, the legacy of formula-based mathematical approximations is still with us in textbooks and software, so we present both in this text.

7.6.7 The Bootstrap Today

The bootstrap was first suggested by Julian Simon in his 1969 work *Basic Research Methods in Social Science* and was named and elaborated by Bradley Efron in the late 1970s. It is now regarded as the major fundamental advance in statistics of the last quarter of the 20th century—a "paradigm shift" in the words of George Casella, chair of the Department of Statistics, University of Florida, writing in

Statistical Science. It now pervades the field of applied statistics, from astronomy to zoology and everything in between. Assessments of uncertainty that were previously infeasible have been made practical by this simple device.

An example is the study of climate data. Manfred Mudelsee, in 2010, tackled the problem of climate data analysis. Uncertainty and error (in the statistical sense) are inevitable in data that purport to represent the state of the climate thousands of years ago. Mudulsee's work looked at nine major time series of climate data, from ice cores in Antarctica to lake sediments in Massachusetts. These time series had been used in climate data models to produce the estimates of global warming and sea level rise.

Mudulsee emphasized that these precise-seeming models are really just estimates and that confidence intervals were essential for a complete understanding of the data. The problem was that the data did not conform to the requirements of the classical statistical methods that had traditionally been used to fit confidence intervals. His work (which you can review at http://bit.ly/1MjRaoE) applied the bootstrap with the goal of establishing credible confidence intervals ("error bars") around various estimates.

7.6.8 Central Limit Theorem

Depending on the size of the sample and the degree of non-Normality in the parent population, sample means are often Normally distributed. This phenomenon is termed the Central Limit Theorem and has been useful historically in facilitating the development of mathematical approximations to sampling distributions.

Definition: Central Limit Theorem The Central Limit Theorem says that the means drawn from multiple samples will be Normally distributed, even if the source population is not Normally distributed, provided that the sample size is large enough and the departure from Normality is not too great.

Many books state that the "large enough" range is a sample size of 20 to 30, but they leave unanswered the question of how non-Normal a population must be for the Central Limit Theorem not to apply.

The Central Limit Theorem allows Normal-approximation formulas to be used in calculating sampling distributions for inference, i.e. confidence intervals and hypothesis tests. With the advent of computer-intensive resampling methods, the Central Limit Theorem is not as important as it used to be, because resampling methods for determining sampling distributions are not sensitive to departures from Normality.

7.7 Standard Error

Another statistic that is used to characterize the reliability of a sample result is the *standard error.*

Definition: Standard Error (SE) The standard error of a sample statistic is the *standard deviation* of that sample statistic. It is often termed the "standard error of the estimate," where "sample estimate" means the same thing as "sample statistic."

Note that the term standard deviation, as used above, refers to variation in the sample statistic, not to variation in the sample data.

Let's illustrate with the results of the Toyota bootstrap, where the sample statistic of interest was the mean. We'll use the 1000 resampled means, as earlier, but instead of finding percentile intervals, we'll calculate the standard deviation of the means.

The estimated standard error is thus 772.4. In calculating the standard deviation of the resampled means, the data are the 1000 means and $n = 1000$. If we had done 10,000 trials, then n would equal 10,000. We divide by $n - 1$ in the calculation, though dividing by n would make little difference.

7.7.1 Standard Error via Formula

The standard error can also be estimated using the standard deviation of the sample data itself, without resampling. Here is the formula for the standard error of the mean, where s refers to the standard deviation of the sample values and n to the sample size.

$$\mathrm{SE}_{\bar{x}} = \frac{s}{\sqrt{n}}$$

7.8 Other Sampling Methods

Simple random sampling is easy in theory, but it can be complex in practice. There are some situations when variations on simple random sampling are either easier to implement or produce better results.

7.8.1 Stratified Sampling

You work for an internet merchant whose websites produce many leads (prospects). Five percent of these leads eventually become sales. You want to conduct a survey to discover what distinguishes non-purchasing leads from those that go on to purchase. In particular, you might want to develop a statistical model that can predict, based on data concerning the prospect, whether a prospect will end up purchasing.

Available to you is a customer database that includes a record for each customer and each prospect. You could take a simple random sample from the database. However, only about 5% of customers in the database will be purchasers. As you proceed with the survey, you will spend time and effort collecting and examining information about lots of non-purchasers while gathering relatively little information about purchasers.

An alternative is to sample equal numbers of non-purchasers and purchasers. This way, you will get the same amount of information about each group.

Definition: Stratified Sampling In stratified sampling, the population is split into categories or strata (singular stratum), and separate samples are drawn from each stratum.

In the above illustration, we need equal information about purchasers and non-purchasers, so we draw equal-sized samples from each stratum. In some cases, it may be desirable to draw different-sized samples from each stratum. For example, those who commission political polls often want to know both about the opinion of the population as a whole as well as the opinions of various subgroups (e.g. Republicans who voted for a Democrat). In such a case, strata sample sizes are a compromise between the need to get information about the population as a whole and the need to get information about subgroups. Taking equal-sized samples of subgroups—particularly smaller groups like Republicans who voted for a Democrat—diminishes the efficiency of the survey for getting information about the population as a whole.

7.8.2 Cluster Sampling

Let's say you want to survey a sample of high school students in a large school district. There is no readily practical and affordable way to achieve the equivalent of placing all the high school students' names in a box and drawing a sample. Various obstacles include the absence of a single student database covering all high schools, privacy concerns, and the manual labor involved in tracking down each student selected for the sample.

An easy and practical solution is to survey the students as a group in their start-of-day administrative homeroom. A certain number of homerooms could be selected from each school, and all the students in each homeroom could be given the survey. This procedure would be reasonably representative, provided that students are assigned to these administrative homeroom units in some fashion that does not introduce bias, such as by birthdate or student ID number. But if students are assigned to homerooms by, for example, athletic participation or academic ability, you run the risk of over- or under-representing groups whose opinions might be systematically different.

Definition: Cluster Sampling In cluster sampling, clusters of subjects or records are selected, and the subjects or records within those clusters are surveyed or measured. The main rationale for selecting clusters, rather than individuals, is practicality and efficiency. Care must be taken that the characteristics that define clusters are not characteristics that will introduce bias into the results.

7.8.3 Systematic Sampling

Another alternative to pure random sampling is systematic sampling—the selection of every nth record. For example, you might sample every 100th transaction or every 10th customer. However, with systematic sampling, you lose the assurance you have with simple random sampling that the sample will be representative of the population—allowing for sampling error. So, you must take care to avoid possible sources of bias. For example, if you sample daily sales and sample every seventh day, you will only measure sales on one particular day of the week. That day may not be typical, so your measurements could be biased.

7.8.4 Multistage Sampling

Professional polling and sampling organizations use a mix of methods to minimize cost, sampling error, and bias. These organizations have the benefit of lengthy experience with different methods. Therefore, they can assess from their prior work whether a departure from simple random sampling, in the name of efficiency, will introduce bias. An example of a multistage process might be to randomly select neighborhoods, as defined by the US Census, and then sample every 10th household.

7.8.5 Convenience Sampling

You have probably encountered polls on the web in which you are invited to state your opinion on something. This is an example of convenience sampling. In convenience sampling, there is no effort to define a population or sampling frame, and there is no attempt to ensure that the sample is representative of a larger population. It is simply taking a sample on the basis of convenience. This sampling method is easy and cheap, but it does not produce consistently useful or reliable information.

7.8.6 Self-selection

An added problem with the convenience sampling method is that the sample is self-selecting. The organization collecting the data is not selecting the sample; the respondents themselves determine whether they participate. In an opinion survey, this almost guarantees biased results—those with stronger opinions are more likely to participate.

The difficulty with self-selection can be seen in web-based product reviews. Consider these two reviews of a printer from Amazon, January 31, 2011.

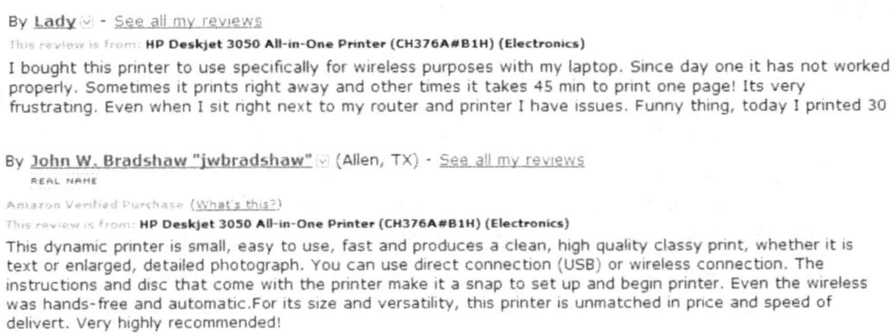
By Lady - See all my reviews
This review is from: **HP Deskjet 3050 All-in-One Printer (CH376A#B1H) (Electronics)**
I bought this printer to use specifically for wireless purposes with my laptop. Since day one it has not worked properly. Sometimes it prints right away and other times it takes 45 min to print one page! Its very frustrating. Even when I sit right next to my router and printer I have issues. Funny thing, today I printed 30

By John W. Bradshaw "jwbradshaw" (Allen, TX) - See all my reviews
REAL NAME
Amazon Verified Purchase (What's this?)
This review is from: **HP Deskjet 3050 All-in-One Printer (CH376A#B1H) (Electronics)**
This dynamic printer is small, easy to use, fast and produces a clean, high quality classy print, whether it is text or enlarged, detailed photograph. You can use direct connection (USB) or wireless connection. The instructions and disc that come with the printer make it a snap to set up and begin printer. Even the wireless was hands-free and automatic.For its size and versatility, this printer is unmatched in price and speed of delivert. Very highly recommended!

Because it appears that anyone who wants to can post a review, the survey method makes it difficult to draw conclusions. Are disgruntled customers more likely to post reviews and skew the results? Does the vendor have an incentive to write spurious positive reviews? What would motivate an ordinary user who had no particular issues to write a review? Are ordinary users who write reviews different from those who do not write reviews?

7.8.7 Nonresponse Bias

Nonresponse bias is a problem that can occur no matter what the sampling method is. It is the question of whether people who respond to a survey are different from those who do not. Since nonresponders, by definition, do not respond and do not show up in surveys, we can't know for sure if they are different from responders. However, it is hard to rule out this possibility.

? How will a survey response rate affect the problem caused by nonresponse bias?

7.9 Absolute vs. Relative Sample Size

Let's return to the story about the Gallup Poll vs. *The Literary Digest*. Gallup believed that selecting a representative sample was more important than selecting a large sample. *The Literary Digest* touted the millions of people in its survey.

Let's consider a related issue. Consider these two questions.

1) What size sample do you need to obtain accurate voter preference estimates for the United States, which has a population of 300 million?
2) What size sample do you need to obtain accurate voter preference estimates for Albany, NY, with a population of 300 thousand?

When asked these questions, many people think that a larger sample is required for the United States, than for Albany. Is this correct?

> **Try It Yourself**
> Suppose a presidential approval survey of 1000 people is conducted for Albany. Spell out the simulation steps, as above, that would be required to assess its accuracy. Then, the same survey is done for the United States as a whole. To achieve the same degree of accuracy for the United States, is a larger sample required?

7.10 Python: Random Sampling Strategies

We already used random sampling with Python at several places in the previous chapters. Here, we will focus on how we can implement the various sampling strategies discussed in this chapter.

7.10.1 Implement Simple Random Sample (SRS)

To set the scene, we implement the sampling experiment from Section 7.3 using Python. The resulting histogram is shown in Figure 7.10.

```
import random
import matplotlib.pyplot as plt

random.seed(123) ①
box = [0] * 128 + [1] * 72 ②
# Repeated sampling
nsamples = 1000
proportion = []
for _ in range(nsamples): ③
```

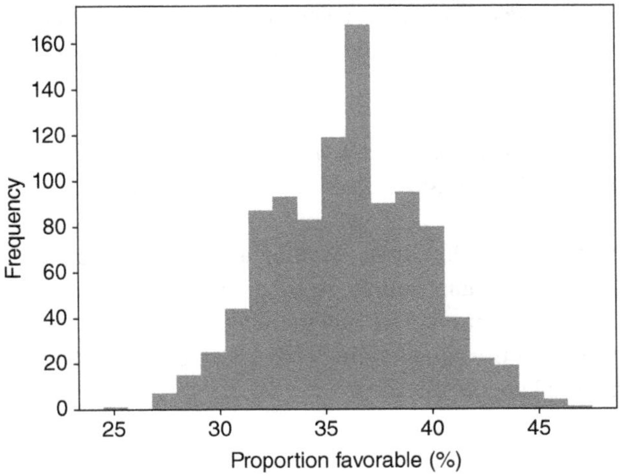

Figure 7.10 Sampling distribution for a proportion.

```
    nr_ones = 0
    for _ in range(len(box)):
        random.shuffle(box) ④
        sample = random.choice(box) ⑤
        if sample == 1:
            nr_ones += 1
    proportion.append(nr_ones / len(box) * 100) ⑥
# Visualize the results
fig, ax = plt.subplots()
ax.hist(proportion, bins=20)
ax.set_xlabel("Proportion favorable [%]")
ax.set_ylabel("Frequency")
plt.show()
print(f"Mean: {sum(proportion)/len(proportion)}")
```

① Set a random seed for reproducibility.

② Create a list with 128 zeros and 72 ones. This represents the box with the 200 cards. Multiplying a list by an integer creates a list with the elements of the original list repeated the specified number of times. This means `[0]*128` will create a list with 128 zeros. Summing two lists concatenates them.

③ The underscore _ is a special variable in Python. It is used to indicate that the variable is not used. In this case, we use because we do not care about the value of the variable.

④ The function *random.shuffle* shuffles the elements of a list in place. This means the list is modified. If you need to preserve the order, create a copy of the original list and work with that.

⑤ The function *random.choice* selects a random element from the shuffled list.
⑥ Convert the number of ones to a proportion [%] and add it to the list of proportions.

You will find out that running this code will take several seconds. Can you spot why this is the case and suggest ways to improve it?

There is no need to shuffle the list 200 times. Making a random choice from the box will be sufficient to get a random sample. Another aspect of writing efficient Python code is to reduce the number of function calls. Instead of calling *random.choice* repeatedly, we can use *random.choices* to get all samples at once. The following code snippet implements this idea.

```
proportion = []
for _ in range(nsamples):
    samples = random.choices(box, k=len(box))
    nr_ones = sum(samples)
    proportion.append(nr_ones / len(box))
# Visualize the results
fig, ax = plt.subplots()
ax.hist(proportion, bins=20)
plt.show()
print(f"Mean: {sum(proportion)/len(proportion)}")
```

The new code is far more efficient and runs in a fraction of a second.

Instead of initializing the box, we can also use the `weights` argument of *random.choices*. In this case, we would write:

```
import numpy as np
population_size = 200
proportion = []
for _ in range(nsamples):
    samples = random.choices([0, 1], weights=[128, 72], k=population_size)
    proportion.append(sum(samples) / population_size)
print(f"Mean: {np.mean(proportion)}")
```

Using the `weights` argument is useful if we know the population size and have information about the distribution. The weights are normalized to sum to 1. This means 128 and 72 are converted to 64% and 36%. Implicitly, the *random.choices* argument converts the weights to `weights=[0.64, 0.36]`.

7.10.2 Determining Confidence Intervals

In Section 7.3.1, we learned how to determine confidence intervals by calculating percentiles from the resampled values. In Python, we can do this as follows:

```
percentiles = [0.025, 0.975] ①
sorted_proportion = sorted(proportion) ②
```

```
lower = sorted_proportion[round(percentiles[0] * nsamples)] ③
upper = sorted_proportion[round(percentiles[1] * nsamples)]
print(f"Confidence interval: [{lower:.3f}, {upper:.3f}]")
```

① Determine the percentiles we are interested in. In this case, we want the 2.5% and 97.5% percentiles, so that we get a 95% confidence interval.
② Sort the list of proportions.
③ Determine the lower and upper bounds of the confidence interval.

Output
```
Confidence interval: [0.295, 0.430]
```

Our calculation is an approximation. We determine the index to lookup the percentile using `proportion[int(p*nsamples)]`. Here, we convert the product `p*nsamples` to an integer so that we can use it to lookup the value in the sorted list. However, what if the product is not a whole number? A more accurate calculation of the percentile is to use, for example, a linear interpolation between the two closest values. The function *np.percentile* implements this and several other methods to determine percentiles; linear interpolation is the default.

```
import numpy as np
lower, upper = np.percentile(proportion, [2.5, 97.5]) ①
print(f"Confidence interval: [{lower:.3f}, {upper:.3f}]")
```

① The *np.percentile* function requires the percentiles to be specified as percentages. The array doesn't need to be sorted.

Output
```
Confidence interval: [0.295, 0.430]
```

We can also use the formula from Section 7.6.2.2 to calculate confidence intervals using a Normal approximation.

```
from scipy import stats
p = np.mean(proportion)
n = population_size
z = stats.norm().ppf(0.975) ①
lower = p - z * np.sqrt(p * (1 - p) / n)
upper = p + z * np.sqrt(p * (1 - p) / n)
print(f"Confidence interval: [{lower:.3f}, {upper:.3f}]")
```

① The *scipy.stats.norm* function returns a Normal distribution with mean 0 and standard deviation 1. The *ppf* method returns the inverse of the cumulative distribution function (CDF) at the specified value. In this case, we want the value at a CDF value of 0.975.

Output
```
Confidence interval: [0.294, 0.427]
```

Alternatively, you can use the `statsmodels` package (more about this package in Chapter 10) to calculate the confidence interval. This package implements a variety of approaches. Here, we show the code for the Normal approximation and for the Clopper–Pearson confidence interval based on the beta distribution.

```
from statsmodels.stats.proportion import proportion_confint
ci_interval = proportion_confint(72, 200, alpha=0.05,
   method="normal") ①
print(f"Confidence interval (normal): [{ci_interval[0]:.3f}, "
      f"{ci_interval[1]:.3f}]")
ci_interval = proportion_confint(72, 200, alpha=0.05,
   method="beta") ②
print(f"Confidence interval (beta): [{ci_interval[0]:.3f}, "
      f"{ci_interval[1]:.3f}]")
```

① The *proportion_confint* function returns a tuple with the lower and upper bounds of the confidence interval. The `alpha` argument specifies the confidence level using the alpha value (1 – 0.95). The `method` argument specifies the method with the default being the Normal approximation.

② With `method="beta"` we select the calculation of the Clopper–Pearson confidence interval.

Output

```
Confidence interval (normal): [0.293, 0.427]
Confidence interval (beta): [0.294, 0.431]
```

7.10.3 Bootstrap Sampling to Determine Confidence Intervals for a Mean

In the previous section, we implemented sampling from a Python list. However, in practice, we will often work with data in a `pandas DataFrame`. The following code snippet shows how we can use bootstrap sampling to determine the confidence interval for the mean price of Toyota Corollas.

```
import pandas as pd
rng = np.random.default_rng(seed=321) ①

df = pd.read_csv("toyota.txt", header=None, names=["price"]) ②

nsamples = 1000
mean_price = []
for _ in range(nsamples):
    sample = df.sample(frac=1.0, replace=True, random_state=rng) ③
    mean_price.append(sample["price"].mean())

percentiles = [0.025, 0.975]
lower, upper = np.percentile(mean_price, [2.5, 97.5])
print(f"Confidence interval: [{lower:.2f}, {upper:.2f}]")
```

① Pandas uses the numpy random number generator. We create a random number generator to ensure reproducibility.
② Read the data from the file toyota.txt into a DataFrame. The file contains a single column with no header. We use the header=None argument to indicate this. We also use the names argument to give the column a name.
③ The *DataFrame.sample* method samples rows from the DataFrame. The argument replace=True indicates that we sample with replacement. (By default, sampling would be without replacement.) There are two ways of specifying the size of the sample. We can use the frac argument to specify the fraction of rows to sample. For bootstrap, we need to create samples of the same size as the original data. Therefore, we use frac=1.0. Alternatively, we can use n=len(df) argument to specify the number of rows in the sample. The argument random_state=rng passes the initialized random number generator to the *DataFrame.sample* method. A simple integer would not work here, because we want to create different samples in each iteration of the loop.

Output
```
Confidence interval: [16242.06, 19112.62]
```

The bootstrap sampling distribution, mean, and 95% confidence intervals are shown in Figure 7.11. The code to create this figure is:

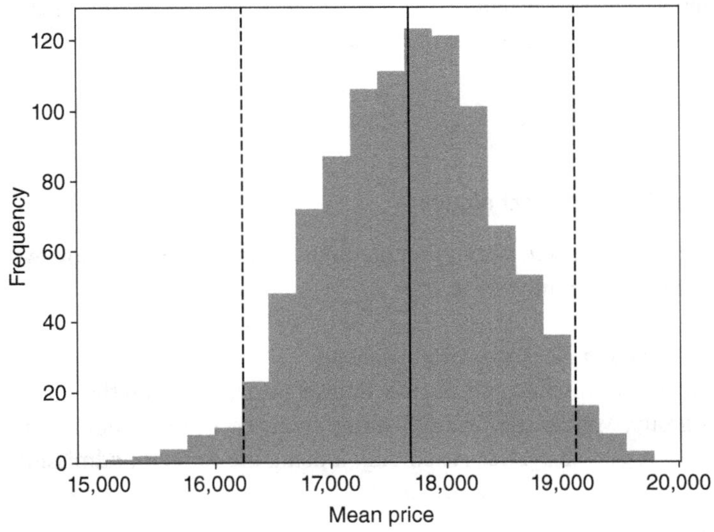

Figure 7.11 Bootstrap sampling distribution for the mean price of Toyota Corollas; mean and 95% confidence interval are indicated by the vertical lines.

```
# Visualize the results
fig, ax = plt.subplots()
ax.hist(mean_price, bins=20)
ax.set_xlabel("Mean price")
ax.set_ylabel("Frequency")
ax.axvline(lower, color="black", linestyle="--")
ax.axvline(upper, color="black", linestyle="--")
ax.axvline(np.mean(mean_price), color="black")
plt.show()
```

We already demonstrated in Section 7.6.5 how we can use the *scipy.stats.t.interval* function to determine the confidence interval via formula, using the *t* statistic. Here is the code for calculating a 95% confidence interval.

```
from scipy import stats

mean_price = np.mean(df["price"]) ①
std_err_mean = stats.sem(df["price"])
dof = len(df["price"]) - 1
# calculate 95% confidence interval
ci_interval = stats.t.interval(0.95, dof, loc=mean_price,
   scale=std_err_mean) ②

print(f'95% confidence interval: [{ci_interval[0]:.2f}, '
      f'{ci_interval[1]:.2f}]')
```

① Calculate the mean and standard error of the mean of the sample.
② Calculate the confidence interval using the *scipy.stats.t.interval* function. The arguments specify the confidence level, degrees of freedom, mean (location), and standard error of the mean (scale).

Output
```
95% confidence interval: [16043.56, 19326.44]
```

7.10.4 Advanced Sampling Techniques

In Section 7.8, we discussed other sampling methods. Here, we will see how we can implement some of them in Python.

7.10.4.1 Stratified Sampling for Categorical Variables

Stratified sampling is done by first splitting the dataset into groups and then sampling from each group. We can achieve this easily with the `pandas` dataframe method `DataFrame.groupby`. Let's exemplify it using the Berkeley admission dataset *microUCBAdmissions.csv*. We use the `Major` column as the group and sample 10% from each group.

```
rng = np.random.default_rng(seed=123)
df = pd.read_csv("microUCBAdmissions.csv")
```

```
resample = (df.groupby("Major") ①
    .sample(frac=0.1, random_state=rng)) ②
print(f"{len(resample)} rows sampled from {len(df)} rows") ③
resample.head()
```

① The *DataFrame.groupby* method groups the rows of the dataframe by the specified column. The *DataFrame.sample* method samples from each group. The `frac` argument specifies the fraction of rows to sample. The `random_state` argument specifies the random number generator to use. The outcome of the *DataFrame.groupby* method is a collection of dataframes.

② The outcome of the *DataFrame.groupby* is *chained* to the *sample* method, which is applied to each group. After sampling from each group, all subsamples are combined in a new `DataFrame` object.

③ The output shows that the new dataframe has 451 rows. This is close to 10% of the original dataset.

Output

```
451 rows sampled from 4526 rows
     Admission  Gender Major
3893  Admitted    Male     A
830   Admitted    Male     A
2916  Rejected    Male     A
925   Admitted    Male     A
844   Admitted  Female     A
```

We can compare the distribution of the departments in the original dataset and the resampled dataset. We see that the distribution is similar.

```
pd.DataFrame({
    "Original": df.value_counts("Major") / len(df), ①
    "Resampled": resample.value_counts("Major") / len(resample)
}).round(3) ②
```

① The *DataFrame.value_counts* method counts the number of rows for each unique value in the specified column. Dividing the counts by the number of rows gives the distribution.

② The *DataFrame.round* method rounds the values to 3 decimal places.

Output

```
       Original  Resampled
Major
A         0.206      0.206
C         0.203      0.204
D         0.175      0.175
F         0.158      0.157
B         0.129      0.129
E         0.129      0.129
```

In the above example, we created a stratified sample using sampling without replacement. By sampling with replacement, we can create a stratified bootstrap sample.

```
rng = np.random.default_rng(seed=123)
resample = (df.groupby(["Major", "Gender"]) ①
    .sample(frac=1, replace=True, random_state=rng) ②
    .reset_index(drop=True)) ③
print(f"{len(resample)} rows sampled from {len(df)} rows")
resample.head()
```

① Grouping can be done on multiple columns by passing a list of column names to the *DataFrame.groupby* method.
② Setting `frac=1` and `replace=True` samples with replacement and will create a bootstrap sample for each group.
③ The *DataFrame.reset_index* method resets the index of the dataframe. The `drop=True` argument indicates that the old index should be dropped. Otherwise, it would be included in the resulting dataframe as a new column.

Output

```
4526 rows sampled from 4526 rows
  Admission  Gender Major
0  Admitted  Female     A
1  Admitted  Female     A
2  Admitted  Female     A
3  Admitted  Female     A
4  Admitted  Female     A
```

Comparing the number of examples with the same gender and major combination, we can see that the resampled dataframe reproduces the distribution of the original dataframe.

```
print(pd.DataFrame({
    "Original": df.value_counts(["Gender", "Major"]), ①
    "Resampled": resample.value_counts(["Gender", "Major"]),
}).sort_index()) ②
```

Output

```
              Original  Resampled
Gender Major
Female A           108        108
       B            25         25
       C           593        593
       D           375        375
       E           393        393
       F           341        341
Male   A           825        825
       B           560        560
       C           325        325
       D           417        417
       E           191        191
       F           373        373
```

7.10.4.2 Stratified Sampling of Continuous Variables

If we want to preserve the distribution of a continuous variable in a stratified sample, grouping by unique values no longer works; we require a different approach. We group the continuous variable into bins and then use the bin as the grouping variable. In the following code example, we demonstrate this using the `CRIM` variable of the *boston-housing.csv*.

```
rng = np.random.default_rng(seed=123)
df = pd.read_csv("boston-housing.csv")

resample = (df.groupby(pd.cut(df["CRIM"], bins=10), ①
                        observed=True) ②
    .sample(frac=0.1, random_state=rng))
print(f"{len(resample)} rows sampled from {len(df)} rows")
resample.head()
```

① The *pd.cut* function creates bins for the specified column. The `bins` argument specifies the number of bins. This command will create bins of equal width. You can also specify the bin edges explicitly using a list. An alternative would be to use the *pd.qcut* function, which creates bins with the same (or almost same) number of elements in each bin.

② The `pandas` package is improving the way it is handling categorical data. We specify the `observed` argument to avoid a warning message. If the continuous variable has larger gaps, it can happen that the binning creates empty bins. The `observed=True` argument of the *groupby* method, will only create groups that are not empty. Setting it to `False` will create all possible groups, even if they are empty.

Output

```
26 rows sampled from 267 rows
        CRIM     RM  MEDV
83   1.38799  5.950  13.2
231  3.69311  6.376  17.7
256  0.14866  6.727  27.5
44   2.73397  5.597  15.4
223  0.63796  6.096  18.2
```

We see the effect of the binning in Figure 7.12. The stratified sample (dashed line) represents the distribution of the full dataset (solid line) more closely than the random sample (dotted line). The figure was created using the following Python code.

```
bw_method = 0.5
ax = df["CRIM"].plot.density(color="black", bw_method=bw_method,
                             label="Full dataset")
resample["CRIM"].plot.density(linestyle="--", ax=ax, bw_method=bw_method,
                              label="Stratified sample")
df["CRIM"].sample(frac=0.1, random_state=rng).plot.density(linestyle=":",
                              ax=ax, bw_method=bw_method,
                              label="Random sample")
```

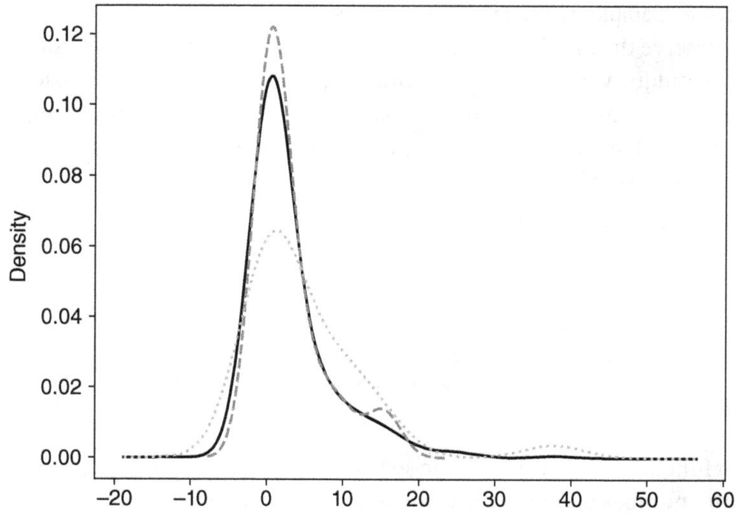

Figure 7.12 Stratified sampling of the `CRIM` variable of the *boston-housing.csv* dataset. The lines show the distribution of the full dataset (solid line), the stratified sample (dashed line), and a random sample (dotted line).

Exercises

7.1 When you take a sample and calculate a statistic (measurement), you learn something about a population—you get an estimate of a population parameter.

a) What is the point of then doing a resampling simulation? What additional information do you get?

b) What about using a formula for a confidence interval? Does its *purpose* differ from using a resampling simulation?

7.2 You work for a survey organization that has been asked to conduct a survey to determine what proportion of football players suffer from health conditions related to head trauma. The client seeking the survey has not been more specific than that. You are developing a plan that can serve as a starting point to discuss exactly what the client wants. Identify:

a) An appropriate population

b) The sampling frame

c) Any other practical issues

7.3 An online retailer is profiling its customers and, among other things, wants to learn its median transaction size for the previous year. Describe a sampling procedure that would avoid having to look at all transactions. (Do not use statistical terms; instead, write down a series of steps that, for example, a summer intern could follow.)

7.4 Consider the following populations:

A = [1, 2, 3, 3]

B = [one million 1's, one million 2's, one million 3's]

a) If you randomly draw one value without replacement from population A, what is the probability of drawing a "3?"

b) If you randomly draw two values without replacement from population A and draw a "3" on your first draw, what is the probability of drawing a "3" on your second draw?

c) If you randomly draw one value without replacement from population B, what is the probability of drawing a "3?"

d) If you randomly draw two values without replacement from population B and draw a "3" on your first draw, what is the probability of drawing a "3" on your second draw?

e) If you randomly draw two values with replacement from population A and draw a "3" on your first draw, what is the probability of drawing a "3" on your second draw?

f) Consider procedures b, d, and e. Which of them are almost equivalent to one another?

7.5 The "churn" rate in a subscription business is the rate at which subscribers leave (cancel) in a given time period. In the US wireless phone industry, the churn rate is close to 10% annually—out of 100 customers at the beginning of a 12-month period, only 90 will be left at the end of the 12 month period.

a) A 10% churn rate can be represented by a single box with 9 cards marked "0" and one card marked "1." Other numbers of 0's and 1's in the box could do the job—give a couple of examples.

b) What resampling process makes these boxes functionally equivalent? Sampling with replacement or sampling without replacement?

c) Describe the resampling steps[2] that you would take to assess the sampling variability of samples of size 250.

d) Describe the resampling steps that you would take to assess the sampling variability of samples of size 50.

e) Implement the resampling steps you described in the previous two questions using Python and compare the results.

2 As in "Put X slips of paper marked "0" and Y slips of paper marked "1" in a hat, take a sample with/out replacement of size …etc.

7.6 A politician contemplating running for local office wants to know how widely her name is recognized and undertakes a sample survey of 200 people who voted in the last election. 65% had heard of him, but the others had not.

a) Describe the resampling steps you would take to determine the reliability of this sample result.
b) Carry out these steps using Python.

7.7 A politician running for state office suggests to the campaign's consultant that, instead of using expensive surveys to measure public opinion among likely voters, they should simply harvest feeds from X (Twitter) and assess sentiment. Comment.

7.8 A lumber mill tests the thickness of plywood panels with a sample of panels. The average thickness in the sample is 0.509 inches and the confidence interval for that mean (based on the sample) is from 0.489 inches to 0.521 inches. Which of the following is true?

a) The observed value is within the interval; therefore, the result is statistically significant.
b) The observed value is within the interval; therefore, the result is not statistically significant.
c) The fact that the observed value lies within the interval is to be expected; this is how confidence intervals work (statistical significance is not at issue).

7.9 Have you ever wondered, when you weigh yourself on a scale, how the scale is calibrated? How is it known that 150 pounds on that scale is the same as 150 pounds on another? Scales are calibrated by comparison to standard weights. Of course, millions of instruments require calibration, and they cannot all be compared to the same weight. Rather, there is a single definitive weight, and a hierarchy of other weights that are compared to it. The US National Bureau of Standards maintains a set of standard weights that are an important link in this chain.

In *Statistics*, Freedman et al. report the results of a set definitive measurements of *one* 10 g standard weight, done in 1962. The first measurement was 9.999591. The error is very slight, on the order of what a fine grain of salt weighs. Several other measurements also came in slightly below 10 g. To make interpretation easier, the Bureau chose to measure not "grams" but rather "micrograms below 10g." A microgram is one-millionth of a gram. So, instead of 9.999591, the measurement was 409. Here is a sample of 20 such measurements for this 10 g standard weight: 409, 400, 406, 399, 402, 406, 401, 403, 401, 403, 398, 403, 407, 402, 401, 399, 400, 401, 405, 402.

a) Based on this sample, you would estimate that this particular standard weight falls short by ___ micrograms.

b) Do a bootstrap simulation to determine how much variability there might be in this estimate; produce a 90% or 95% confidence interval.

7.10 The data *streams.csv* reflects the pH readings on a sample of streams in a United States county.

a) Calculate a point estimate and a confidence interval (90% or 95%) for the mean pH.

b) Discuss briefly a possible sampling scheme that might have been used to obtain these data, and potential challenges and pitfalls.

7.11 Using the *pulse.csv* data:

a) Calculate the proportion of the total who smoke and find a confidence interval (90% or 95%) around this proportion.

b) It appears that males are more likely to be smokers than females. Find the difference in the proportion who smoke (males–females) and find a confidence interval (90% or 95%) around that difference.

c) Medical theory suggests that smoking elevates the pulse rate. Looking at the "before" pulse rate for everyone, calculate the difference: "mean smoker pulse rate" minus "mean nonsmoker pulse rate." Find a confidence interval around this difference.

8

More than Two Samples or Categories

In previous chapters, we looked at the popular A/B test and how to test for statistical significance of the results. In this chapter, we will see how to compare multiple samples and whether they differ from one another. After completing this chapter, you should be able to:

- Construct an R × C table to compare categorical data across multiple categories.
- Perform a chi-square test to assess whether categories differ in a statistically significant way.
- Construct a table to compare numeric data across multiple samples or treatments.
- Conduct ANOVA to test whether multiple samples with numeric data differ in a statistically significant way.
- Explain how the problem of testing for statistical significance becomes more complex with multiple comparisons.
- Explain how multi-arm bandits deal with A/B and multiple testing from an optimization perspective.

8.1 Count Data—R × C Tables

The field of behavioral economics has called into question many truisms of classical economics, such as the idea that demand rises as prices go down. The business consultant McKinsey and Company recounts the story of a jewelry store owner trying to sell a line of jewelry that was proving difficult to move.

Several strategies accomplished nothing, and the owner asked the staff to mark down the price by "X1/2" and left on a trip. On her return, she was surprised to see the whole lot gone and doubly surprised to learn that the staff had misinterpreted instructions and *doubled* the prices instead of discounting them. Academics studying consumer behavior have learned that "second most expensive" and "second cheapest" are good price points that boost demand irrespective of product

Statistics for Data Science and Analytics, First Edition. Peter C. Bruce, Peter Gedeck, and Janet Dobbins.

Companion website: www.wiley.com/go/Wiley_Statistics_for_Data

Doubling the price of some jewelry increased its popularity.
Source: PhotoMIX Company/Pexels.

quality. Purchasers at the top end tell themselves they are not paying absolute top dollar, and those at the bottom can avoid thinking of themselves as cheap. Adding a couple of relatively inferior items alongside a flagship item can boost sales of the latter, as it makes the choice easier. On the other hand, including too many items can induce decision paralysis.

8.2 The Role of Experiments (Many Are Costly)

Knowledge of human behavior is useful in business, but much of this knowledge is not obvious and must be derived through experiments. The digital age has made some marketing experiments cheap and relatively easy, but behavioral experiments can be expensive to set up. Industrial and medical studies, as well as surveys, are even more expensive.

Clinical trials for new drugs can cost hundreds of millions of dollars. The overall design is set up front and cannot be modified midstream. Once the data is collected, which can take years in medical experiments, the results are what they are. It is not possible to go back and revise the design. For this reason, it often makes sense to investigate several things at once.

Let's look first at an example in behavioral psychology.

8.2.1 Example: Marriage Therapy

Do behavioral and insight therapies for marriage counseling differ in effectiveness? Behavioral therapy stresses the skills of managing interpersonal

Table 8.1 Marriage therapy.

	Happily Married	Distress Married	Divorced
Behavioral	15	3	11
Insight	24	5	1

Source: Adapted from Douglas et al., 1991.

relationships, and insight therapy stresses working out underlying difficulties. Fifty-nine couples were randomly assigned, with 29 to behavioral therapy and 30 to insight therapy. At a four-year follow-up, 15 of the behavioral group were happily married (HM), three were distressed married (DM), and 11 were divorced (DIV). The insight group had 24 HM, five DM, and one DIV (see the contingency table for these results, Table 8.1).

Are the differences among the groups significant?

We could ask whether insight therapy produces more "happily married" results, but that leaves out the "distressed married." And which is better, "distressed married" or "divorced"? Even if we could say which is better, we still do not have a simple comparison of the two groups that can be boiled down to a single statistic.

A single statistic is necessary for a hypothesis test, which in the end will have to answer the question "might this have happened by chance?" where "this" is the extreme value of some statistic.

So, we can step back and ask a more general question: Do the observed results depart from what we would expect to get by chance if the choice of therapy had no effect at all on the outcome?

Departure from Expectation

The concept of "departure from expectation" is an important one in statistical inference. By this, we really mean "departure from what we would expect if only chance variation were at work."

We have already been testing hypotheses to see if a mean or proportion is bigger (smaller) in one sample than another. Departure from expectation is a more general overlapping concept that simply asks whether observed sample results (from two or more samples) are *different* from what we would expect in a chance model.

The fundamental idea is to figure out what a sample result (e.g. proportion or mean) would be in a chance model, what it was in actuality, and subtract. The concept can be applied to both continuous (measured) data and count data.

For continuous data, the traditional approach to measuring departure from expectation is *Analysis of Variance* (ANOVA), which is dealt with later in this chapter.

For count data, the traditional approach is a *chi-square test*, which we take up next.

8.3 Chi-Square Test

Before we revisit the chi-square test, which we introduced in Chapter 6, we will construct a resampling test with an intuitive statistic—sum-of-differences. The first task is to establish what we would expect if both therapies yielded the same results, i.e. our null hypothesis is that results are independent of therapy. We would expect the overall behavioral/insight split to be the same across all marital outcomes. With 29 in the behavioral group and 30 in the insight group, if therapy made no difference, we would expect each outcome (happily married, distressed, divorced) to be almost equally split between behavioral and insight therapy (with a bit more to the insight group).

There were 39 happily marrieds. Specifically, we would expect just under half, or 29/59 = 49.2% = 19.17 couples, to be behavioral and just over half of them, or 30/59 = 50.8% = 19.83 couples, to be insight. We can make similar calculations for distressed married and divorced.

8.3.1 Alternate Option

With equivalent arithmetic, you could note that there were 39 HMs (66.1%), 8 DMs (13.6%), and 12 DIVs (20.3%). So, of the 29 behavioral couples, 66.1% = 19.17 couples would be expected to be HM and so on (Table 8.2).

The next step is to determine the extent to which the observed results differ from the expected results. Direction does not matter, so we take absolute values (Table 8.3).

Overall, these differences sum to 20.41, although the table above sums to 20.42 due to rounding.

8.3.2 Testing for the Role of Chance

Is this a greater sum of differences than we might expect from a random allocation of outcomes to 29 behavioral couples and 30 insight couples?

To answer this question, we can use the following resampling procedure (also called a permutation procedure).

Table 8.2 Expected outcomes if treatments yield the same results.

	Happily Married	Distress Married	Divorced
Behavioral	19.17	3.93	5.90
Insight	19.83	4.07	6.10

Table 8.3 Absolute difference between observed and expected.

	Happily Married	Distress Married	Divorced
Behavioral	4.17	0.93	5.11
Insight	4.17	0.93	5.11

1) Fill a single box with 39 ones (happily married), 8 twos (distressed married), and 12 threes (divorced).
2) Shuffle the box and take two samples without replacement of sizes 29 and 30.
3) Count the number of ones, twos, and threes in each sample.
4) Reconstruct the resampled counterpart to Table 8.1.
5) Find the sum of absolute differences between that resampled table and what you would expect under the null hypothesis (Table 8.2).
6) Repeat steps two through five many times, say 10,000 times, each time recording the sum of absolute differences.
7) Determine how often the resampled sum of differences exceeds the observed value of 20.41.

The outcome of one resampling trial can be seen in Table 8.4; we shuffle the box and then treat the first 29 values as the $N = 29$ resample and the remaining 30 as the $N = 30$ resample.

The statistic of interest here was 4.27 (Table 8.5), which is not nearly as extreme as the observed value of 20.42. Of course, the above is just one trial. When the simulation was repeated 10,000 times, a statistic $\geq$20.42 was encountered, but only 65 times.

A portion of the relevant resampled values in the vicinity of the observed value of 20.42 is displayed in Table 8.6 from the sorted output of 10,000 trials.

We can see that only 65 trials (one through 65) out of all 10,000 trials—fewer than 1%—yielded a sum of differences as big as or bigger than the observed value.

Table 8.4 Resampling output—distribution of outcome after one shuffling.

Therapy	Outcome	Therapy	Outcome
Behavioral	Divorced	Insight	Happily married
Behavioral	Happily married	Insight	Happily married
Behavioral	Divorced	Insight	Happily married
Behavioral	Happily married	Insight	Happily married
Behavioral	Happily married	Insight	Happily married
Behavioral	Happily married	Insight	Divorced
Behavioral	Happily married	Insight	Divorced
Behavioral	Happily married	Insight	Happily married
Behavioral	Happily married	Insight	Happily married
Behavioral	Happily married	Insight	Distress married
Behavioral	Happily married	Insight	Happily married
Behavioral	Divorced	Insight	Distress married
Behavioral	Distress married	Insight	Divorced
Behavioral	Happily married	Insight	Distress married
Behavioral	Happily married	Insight	Happily married
Behavioral	Happily married	Insight	Happily married
Behavioral	Happily married	Insight	Divorced
Behavioral	Distress married	Insight	Divorced
Behavioral	Distress married	Insight	Happily married
Behavioral	Happily married	Insight	Happily married
Behavioral	Happily married	Insight	Happily married
Behavioral	Happily married	Insight	Happily married
Behavioral	Happily married	Insight	Happily married
Behavioral	Divorced	Insight	Happily married
Behavioral	Happily married	Insight	Divorced
Behavioral	Distress married	Insight	Happily married
Behavioral	Divorced	Insight	Divorced
Behavioral	Happily married	Insight	Happily married
Behavioral	Distress married	Insight	Happily married
		Insight	Happily married

Table 8.5 Tabulation of one shuffling.

Resampled table

	Happily Married	Distress Married	Divorced
Behavioral	19	5	5
Insight	20	3	7

Expected table

	Happily Married	Distress Married	Divorced
Behavioral	19.17	3.93	5.90
Insight	19.83	4.07	6.10

Absolute differences

	Happily Married	Distress Married	Divorced
Behavioral	0.17	1.07	0.90
Insight	0.17	1.07	0.90

Sum = 4.27

This represents an estimated *p*-value of 0.0065. The chance occurrence of the observed value is so rare that we conclude that there is a real difference among the therapies.

8.3.3 Standardization to the Chi-Square Statistic

With computer-based resampling, the use of absolute values is no problem, and we can compare the observed sum-of-difference statistic to values obtained by the resampled distributions. Before computing power was widely available to do your own resampling test, the chi-square statistic was developed to allow comparison of *any* 2×2 table to a standardized and tabulated test statistic. That same statistic can be generalized to the case of an $R \times C$ table.

Definition: Chi-Square Statistic (repeating from Chapter 6) The chi-square statistic squares each deviation from expectation in a contingency table, divides it by the expected value, and then sums the results.

Table 8.6 Marriage therapy: A few of the sums of resampled differences, in descending order out of 10,000, in the vicinity of the observed value.

62	20.678
63	20.678
64	20.678
65	20.678
66	20.407
67	20.407
68	20.407
69	19.593
70	19.593
71	19.593

Dividing by the expected value standardizes the distribution so that it is a member of a family of similar chi-square distributions, each characterized by its degrees of freedom. For an R × C (rows by columns) table, the degrees of freedom are calculated as (R-1)(C-1). This standardization allows a comparison between the results of a given study like the one above and a tabulated set of chi-square values.

8.3.4 Chi-Square Example on the Computer

Most statistical software has the ability to perform chi-square tests. Here is an example using Python:

```
result = stats.chi2_contingency(observed)
print(f"chi2 = {result.statistic:.3f}")
print(f"p-value = {result.pvalue:.4f}")
print(f"degrees of freedom = {result.dof}")
print("expected")
print(result.expected_freq)
```

The command creates this output:

```
chi2 = 10.896
p-value = 0.0043
degrees of freedom = 2
expected
[[19.16949153  3.93220339  5.89830508]
[19.83050847  4.06779661  6.10169492]]
```

8.4 Single Sample–Goodness-of-Fit

The idea of departure-from-expectation can also be applied to a single sample. In 1991, Tufts University researcher Thereza Imanishi-Kari was accused of fabricating data in her genetic research on mice. Congressman John Dingell became involved, and the case eventually led to the resignation of her colleague, David Baltimore, from the presidency of Rockefeller University.

Imanishi-Kari was ultimately exonerated after a lengthy proceeding. However, one element in the case rested on statistical evidence regarding the distribution of digits in her data, as compared to the expected distribution of interior digits in data produced by a wide variety of sources.

Leading Digits: Looking at data from many different natural and man-made sources, it turns out that "1" occurs as the initial digit in numbers almost twice as frequently as the next most frequent leading digit, "2." Digits "3" through "9" decline steadily in frequency as the leading digit. See Figure 8.1 and Google "Benford's Law" for more information.

Interior Digits: Benford's Law applies to leading digits, but not interior digits in long numbers—they are expected to show a uniform distribution, with each one occurring with probability 1/10. Imanishi-Kari's data were examined, and the investigator concluded that the distribution of her interior digits strayed too far from the expected uniform distribution.

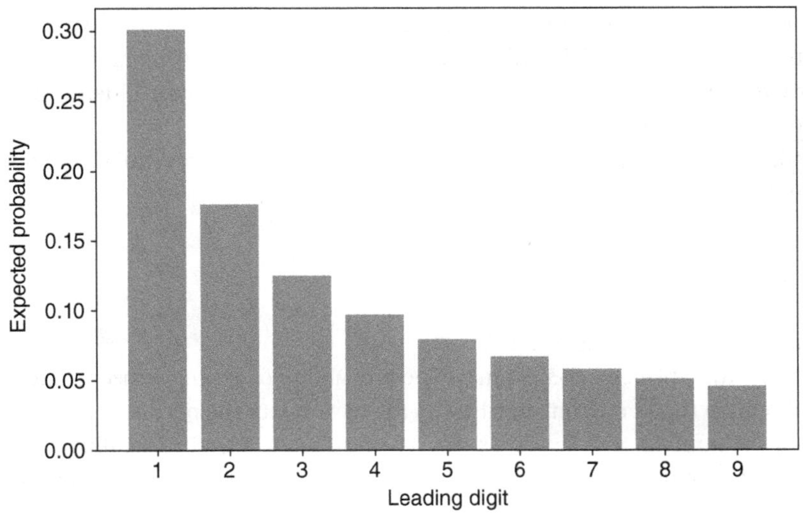

Figure 8.1 The frequency (*y*-axis) of leading digits (*x*-axis) in most multi-digit numbers that count or measure something.

Table 8.7 Frequencies of 315 interior digits in Imanishi-Kari data.

Digit	Frequency
0	14
1	71
2	7
3	65
4	23
5	19
6	12
7	45
8	53
9	6

Source: *Science*, March 8, 1991, News & Comment, p. 1171.

What is too far? Table 8.7 displays the frequencies of 315 interior digits in one of Imanishi-Kari's data tables.

The distribution looks unbalanced, but could the imbalance arise naturally through random chance? Let's test the proposition that the digits 0 through 9 all occur with equal probability. Note how we measure the difference between the resampled distribution and the expected uniform distribution, i.e. the expected distribution is that each digit will occur 31.5 (315/10) times. In Imanishi Kari's table, we got fourteen zeros when we expected 31.5, so the absolute difference is 17.5. For the ones, the absolute difference is 71 − 31.5 or 39.5.

Overall, these absolute deviations sum to 216. Is this a greater overall deviation than might be explained by chance?

8.4.1 Resampling Procedure

1) Create a box with the digits 0, 1, 2, … 9.
2) Sample with replacement 315 times. The probability of getting a specific digit must remain the same from one draw to the next, hence the need to replace it.
3) Count the number of 0s, 1s, 2s, 3s, etc. The histogram in Figure 8.2 illustrates just one group of 315 digits using a bin width of one. It looks more balanced than Imanishi-Kari's.
4) Find the absolute difference between the number of 0s and 31.5, the number of 1s and 31.5, etc., and sum.

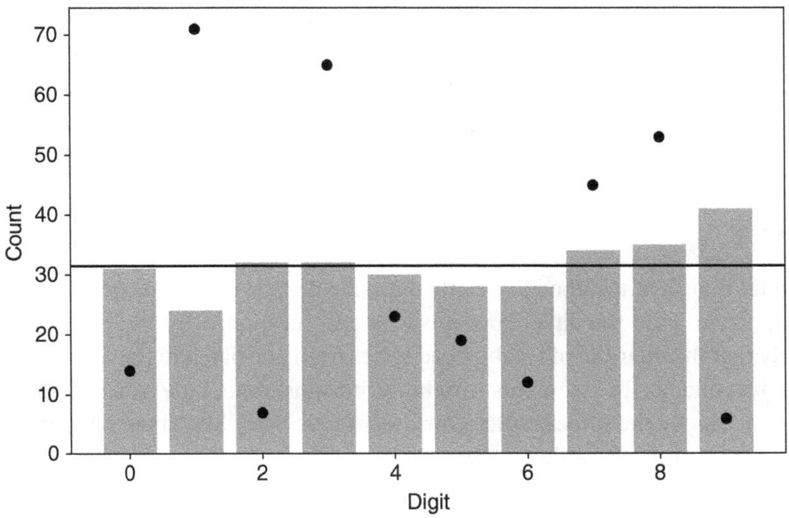

Figure 8.2 Histogram of one 315-digit resample (bars) compared to the observed distribution of interior digits in Imanishi-Kari's data (dots).

5) Repeat steps 2–4, 10,000 times.
6) How often did the resampled sum of absolute deviations equal or exceed 216? Divide this sum by 10,000 to calculate the *p*-value.

Results: Not once in 10,000 trials did an imbalance as great as the observed imbalance occur, for an estimated *p*-value = 0.0000.

These results show that an imbalance as great as the observed one is extremely rare, so we reject the null hypothesis that chance is responsible. This does not prove that Imanishi Kari invented the results; it is possible that some other non-chance mechanism could be at work.

This statistical procedure is known as a "goodness-of-fit" test. It examines how well an observed distribution fits a theoretical expectation.

8.5 Numeric Data: ANOVA

We turn now from count data to multiple groups with numeric (measured) data. Although the data below dates back to 1935, the topic of the study—the amount of fat in our diet—is still a very current issue. In this study, investigators wanted to know how much fat doughnuts absorb while they are being fried. In particular, they wished to compare the absorption levels of four types of fat.

The investigators whipped up a batch of doughnut dough, split it into four doughnuts, and fried one doughnut in each type of oil (or fat). The results are given below.

- Fat 1 164 g
- Fat 2 178 g
- Fat 3 175 g
- Fat 4 155 g

If each fat is always absorbed to exactly the same degree, and there are no measurement errors, then the data above answers our question. Unfortunately, it is very likely that the quantity of fat absorbed varies from one batch to the next, even if we use just one fat. To assess this variability, we must repeat the experiment to see if we get exactly the same results. Our investigators did this. The second time around, they obtained the following results:

- Fat 1 172 g absorbed
- Fat 2 191 g absorbed
- Fat 3 193 g absorbed
- Fat 4 166 g absorbed

The results above are clearly not the same numbers when compared to what the investigators observed the first time around. For example, during the second experiment, Fat 3 was absorbed the most, whereas in the first experiment, Fat 2 was absorbed the most. A repetition of an experiment that allows us to assess variability is called a "replication."

Definition: Replication A replication is a repeat of an experiment or a procedure with all elements remaining unchanged.

Actually, our investigators made six replications using just one large batch of dough.

What are the advantages and disadvantages of using one big batch of dough rather than a separate batch for each replication?

In at least one replication, Fats 1, 2, and 3 each had, at one point, the highest levels of absorption. There were also replications in which Fats 1 and 4 had the lowest levels of absorption. Now, our results are not so clear-cut. Let's look at the complete data set in Table 8.8.

Of course, it is always important to graph our data in addition to examining summary statistics. Figure 8.3 shows dot plots of the data (a dot plot is used with

Table 8.8 Fat absorption data (grams).

	Fat 1	Fat 2	Fat 3	Fat 4
Replication 1	164	178	175	155
Replication 2	172	191	193	166
Replication 3	168	197	178	149
Replication 4	177	182	171	164
Replication 5	156	185	163	170
Replication 6	195	177	176	168

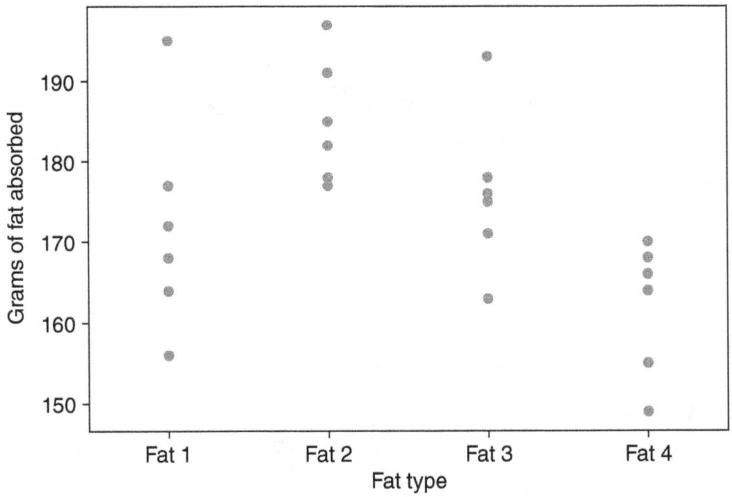

Figure 8.3 Dot plots for the doughnut experiment.

relatively small datasets; each observation gets its own dot). Figure 8.4 shows boxplots of the data.

It looks like the groups differ, but it also looks like there is a lot of overlap among the groups. We need some criterion for deciding just how different the samples must be before we say there is a real difference in the population. In other words, we need some sort of hypothesis test.

What statistic will we use in our hypothesis test? There are several choices, but the average amount of fat absorbed is probably of primary interest, so we will compare means, as shown in Table 8.9.

Now, we have a conundrum. When we were comparing just two groups, it was a simple matter; we merely looked at the difference between the means of each group. With four means, there are six possible comparisons between fat types.

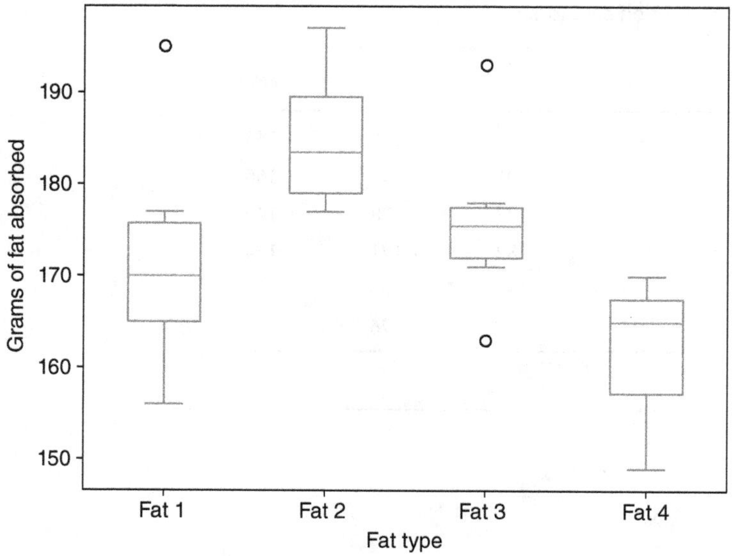

Figure 8.4 Boxplots for the doughnut experiment.

Table 8.9 Comparing means.

	Fat 1	Fat 2	Fat 3	Fat 4
Replication 1	164	178	175	155
Replication 2	172	191	193	166
Replication 3	168	197	178	149
Replication 4	177	182	171	164
Replication 5	156	185	163	170
Replication 6	195	177	176	168
Average	172	185	176	162

- Fat 1 compared to Fat 2
- Fat 1 compared to Fat 3
- Fat 1 compared to Fat 4
- Fat 2 compared to Fat 3
- Fat 2 compared to Fat 4
- Fat 3 compared to Fat 4

We could do multiple hypothesis tests, but that would increase our chances of being fooled by chance (see Section 8.8). Instead of conducting multiple tests, we will examine multiple groups with a single test, which is called an analysis of variance (ANOVA). If the results of this test are significant, this means there is some significant difference among the group means as a whole. Our null hypothesis is that all four fats share the same absorption properties, and the differences among the groups are just due to chance.

We can test this hypothesis by repeatedly placing all the values in a box, randomly drawing out four groups of six, and then measuring the extent to which the group means differ from the overall "grand mean." There are different ways to measure variation, and we will choose the variance using the following formula (k refers to the total number of groups, and $k - 1$ is the counterpart to $n - 1$ used in calculating variance for individual observations):

$$\text{Variance among group means} = \frac{\sum (\text{grand average} - \bar{x})^2}{k - 1}$$

The observed variance among the group means is 90.92. Table 8.10 shows the variances among group means for the experiment.

"Deviation" in the above table refers to the difference between the group mean and the grand average. It is also the "group effect" or "treatment effect."

The steps in our resampling procedure are as follows:

1) Place all 24 fat absorption values in a box.
2) Shuffle the box and draw[1] four groups of six values each (i.e. four groups, each with six values).
3) Find the mean of each of the four groups.
4) Calculate the variance among the means and record it.

Table 8.10 Variance of group means.

Fat Type	Group Mean	Deviation	Dev. Sq.
1	172	−1.75	3.06
2	185	11.25	126.56
3	176	2.25	5.06
4	162	−11.75	138.06

Grand mean 173.75 Total 272.75

Variance = 272.75/(k − 1) = 272.75/3 = 90.92

1 We draw without replacement here. Where you have combined two or more groups in a single box, drawing from that box can be done with or without replacement. The conceptual frameworks and statistical properties differ a bit in ways that are beyond the scope of this text, but both are valid procedures.

Table 8.11 Resampled variances.

128.602
113.639
111.954
104.750
100.750
98.157
91.046
Threshold
90.602
87.361
86.491

5) Repeat steps 2 through 4 many more times (say up to 1000 times).
6) Determine what proportion of the 1000 trials produces a variance ≥90.92. This is the *p*-value.

If the trials regularly produce a variance in excess of the observed value of 90.92, this indicates that chance is a reasonable explanation for the differences among the groups.

The largest 10 results out of 10,000 trials produced by this resampling procedure, and ordered from largest to smallest, are given in Table 8.11.

You can see that only seven of the 1000 trials produced a variance as great as 90.92, for an estimated *p*-value of 0.007. We, therefore, conclude that chance is probably not responsible for the differences among the fats with respect to absorption in the doughnuts, and those differences are statistically significant.

As shown in Figure 8.5, the distribution of the variances of the 1000 trials is skewed to the right and truncated at zero. Variances less than zero are not arithmetically possible.

8.6 Components of Variance

In the doughnut problem, each observed data value in the data set can be thought of as a combination of three components:

Observed value = Grand average + Treatment effect + Residual error

Sometimes, this ANOVA model is written in symbols rather than in words:

$$Y_{ij} = \mu + \alpha_i + \epsilon_{ij}$$

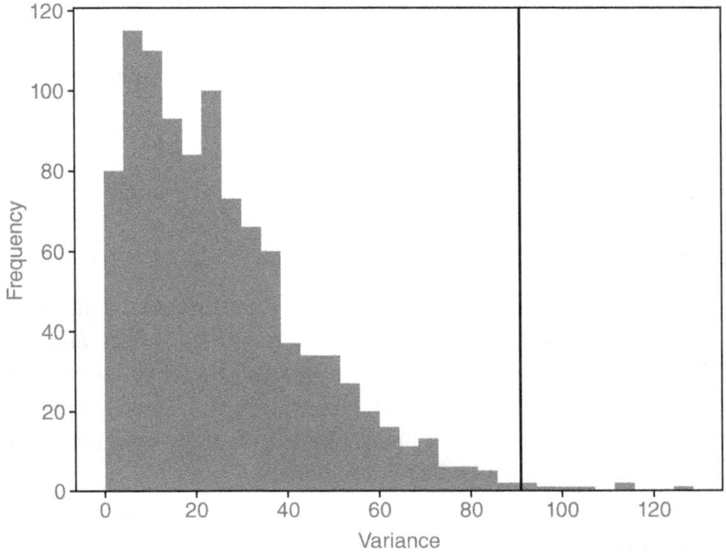

Figure 8.5 Frequency histogram of resampled variances from doughnut data problem.

In the above model, Y_{ij} is used to denote the individual observation. The Grand Average is denoted as μ, the treatment effect as α_i, and the residual error as ϵ_{ij}.

You can see this laid out in Table 8.12, along with the means for each fat and the grand average.

One goal of ANOVA is to estimate the amount of variability in our dependent (response) variable that is due to the independent predictor (treatment) variable, as opposed to variability due to individual differences among cases.

Table 8.12 Doughnut data with group means and grand average.

Fat 1	Fat 2	Fat 3	Fat 4
164	178	175	155
172	191	193	166
168	197	178	149
177	182	171	164
156	185	163	170
195	177	176	168
Average $\bar{x}_1 = 172$	Average $\bar{x}_2 = 185$	Average $\bar{x}_3 = 176$	Average $\bar{x}_4 = 162$
			Grand average = 173.75

The leftover variability *not* attributed to the independent variable is referred to as residual error.

The statistician George Cobb suggested an "assembly line" metaphor to help explain the above model.

- Start with the grand average (173.75 for doughnut data)
- Add treatment or predictor variable effect, remembering that it might be negative (independent variable = fat type)
- Add residual error, which also might be negative.

By an analogous reverse process, we can "decompose" any observed data value within a data set into the grand average, the treatment or variable effect, and the residual error. The ratio between the treatment effect and the residual error is important—the higher this ratio is, the more the treatment effect stands out from the "noise" of residual error.[2] For more on this, including the traditional metric of this ratio, the F-statistic, see the Appendix to this chapter.

8.6.1 From ANOVA to Regression

In ANOVA we are interested in *whether* there is a statistically significant difference among groups. We might also be interested in this question, "How *much* does the outcome change as you change the input variables?" We will take up this issue in Chapter 10.

8.7 Factorial Design

The type of experiment we have been reviewing—an ANOVA with multiple independent variables whose levels are set by the researcher—is called a *factorial design*. To see the advantages of factorial design, let's look at a hypothetical situation and consider other ways a study might be conducted. The hypothetical situation is that there are two types of subjects, A and B (for example, male and female), and two treatments, 1 and 2. A factorial design produces a table like Table 8.13 of mean outcomes for the four factor combinations.

Clearly, there is a very strong interaction between treatment and subject type, meaning that the effect of one independent variable on the response depends on the level of the other independent variable. The outcomes differ markedly for subject types A and B. Now consider other possible designs and how they would turn out.

---

2 The term "noise" implies randomness, while "residual error" could include something other than random variation, such as the effect of some variable not considered in the analysis. Often, the two terms are used interchangeably.

Table 8.13 Hypothetical outcomes of factorial design.

	Treatment 1	Treatment 2
Subject type A	10	20
Subject type B	15	5

?

1) One investigator recruits only subject type A. What will her investigation conclude about the effect of treatments one and two?
2) A second investigator recruits only subject type B. What will her investigation conclude about the effect of treatments one and two?
3) A third investigator recruits both subject types but keeps no record about which subject types get treatment one and which get treatment two. She has only the overall results. What will her investigation conclude about the effect of treatments one and two?

The factorial design allows us to understand how effects differ for different groups and avoid the above errors.

8.7.1 Stratification and Blocking

In an experiment, the researcher can often control the levels of all the variables. In survey sampling, a researcher may not be able to alter a value for a subject, but he or she may be able to choose subjects with appropriate values. Imagine a survey to be done in Vermont, for which it is believed race or ethnicity may be an important variable. We cannot assign a race or ethnicity to a subject. The demographics of Vermont are such that, in a general sample, there are apt to be many whites but very few blacks, native Americans, or Hispanics.

To overcome that problem, we can draw four separate samples from the four ethnic groups instead of from the population as a whole. In that case, the subpopulations are called strata, and the process is called stratification. It is limited in that it may not be practical to stratify on all variables at once if there are many of them. Fortunately, we can often pick and choose which variables we wish to stratify. Gender might also be a variable in the same survey, but the numbers of males and females in Vermont are balanced closely enough that stratification would not be needed. Stratification is needed, in other words, when the groups we need to sample make up a small part of the general population.

The strata then become variables. In the Vermont case, we might have a single categorical variable "stratum" that can take on the values, "White, Black, Native American, Hispanic."

To gain an accurate picture of the whole population, of course, you cannot simply combine the results from the different strata. You must weight them by their share of the population.

8.7.2 Blocking

Blocking: Stratification is typically applied to observational or survey data. In designed experiments, the equivalent concept is known as "blocking."

Often, the blocking variable, or the variable on which we stratify, is not one of the variables of interest in an experiment but rather a "nuisance" variable. Such a nuisance variable is one we can measure and include in the study but whose effect we are not interested in studying. For example, an internet retailer may be interested in testing the implementation of new AI in its phone support system and chooses several call centers to test the new system. It would include the call center identity as a blocking variable because effects may differ from center to center, even though the company is not currently interested in studying differences among call centers.

8.8 The Problem of Multiple Inference

One common abuse of data analysis is to gather or find data and then look up and down and left and right in an attempt to find some association, pattern, or difference between groups that is interesting. The more questions you investigate, the greater the chances you will find something "significant." The related sin in hypothesis testing is to gather data first and then test all kinds of hypotheses until we find one that gives a significant result. We often refer to this sin as *data snooping*.

One option we have at this point, with our doughnut data, is to simply conduct six different A/B tests, as you learned about in Chapter 2. This would allow us to examine the six comparisons between fat types that were outlined at the end of Section 8.5. Intuitively, it might make sense to you to do this, but there are some problems with this approach. Because each test or comparison will falsely show a significant result 5% of the time, you can usually find such an erroneous result if you make enough comparisons; in fact, about one in every 20 comparisons you make will mistakenly appear significant. Put more simply, the more comparisons we make, the more prone we are to make an error and to claim that a comparison is significant when it really is not.

This is not simply a technical issue. It is a fundamental problem with data analysis as it is practiced today, particularly in the health arena, and there are significant costs.

Everything is Dangerous

"Everything is Dangerous" was the title of a lecture that statistician Stanley Young gave at a number of US government agencies in the first decade of the 21st century. He was referring to published reports about associations between certain things and human health, such as coffee and pancreatic cancer, type A personality and heart attacks, reserpine and breast cancer, etc.

Young and a colleague, Alan Karr, expanded on this theme in a September 2011 article in *Significance* magazine. They examined 52 claims of health benefits reported in 12 peer-reviewed papers. These papers all reported observational studies rather than experiments (i.e. they were reviews of existing data). One of the observational studies claimed protection from heart attack afforded by vitamin E, which we mentioned at the beginning of this book.

Of course, there are hundreds—thousands—of claims every year about substances or treatments represented as helpful or harmful to health. What was special about these 52 claims was that they were later tested in rigorously controlled randomized trials. Amazingly, *not one* of the 52 claims was verified, and the trials produced opposite results in five of the cases.

The most likely reason for the failure to validate these claims in experiments is that they were the result of data snooping in the first place. Data snooping is an extensive search through data, testing a variety of possibilities until something interesting is found. Other phrases describing this effect include:

- Torturing the data until it confesses (Ronald Coase)
- The vast search effect (John Elder)

A conservative approach that avoids such problems is to state any hypotheses to be tested in advance. If a researcher cannot state a definite hypothesis, then it is probably too early in the study to test anything, and more data is needed to develop at least one hypothesis.

Examining data in order to develop a hypothesis is termed *exploratory data analysis (EDA)*. These techniques for summarizing and visualizing data were described in Chapter 3. Once you have developed a hypothesis and want to test it for statistical significance, you should use *different* data. Significance tests on the same data whose exploration suggested the hypothesis are biased in favor of finding significance.

If you have more than one hypothesis to test, then you must adopt a more stringent standard for any one of them to be declared "significant." A simple, though conservative, approach is to divide the desired alpha by the number of tests. For example, if you have five hypotheses to test, and the traditional alpha (i.e. significance level) you require for a single hypothesis test is 0.05, then the equivalent alpha for each of five hypothesis tests is 0.01 (or 0.05/5). There are more complex schemes for alpha adjustment beyond the scope of this book.

8.9 Continuous Testing

To this point, our process has been that data are collected in an experiment, according to a specified design, including sample size, to answer a specific question, e.g. "Which is better, treatment A or treatment B?" The presumption is that once we get an answer to that question, the experimenting is over, and we proceed to act on the results. You can probably perceive several difficulties with that approach.

1) Our answer may be inconclusive: "effect not proven." The results from the experiment suggest an effect, but we don't have a big enough sample to prove it.
2) We might want to begin taking advantage of results that come in prior to the conclusion of the experiment.
3) We might want the right to change our minds or try something different based on additional data that comes in after the experiment is over.

The traditional approach to experiments and hypothesis tests dates from the 1920s and is rather inflexible. The advent of computer power and software has made more flexible approaches possible, particularly in business but in medicine as well.

8.9.1 Medicine

The strictest rules governing experiments apply, naturally, in medicine and, in particular, in drug trials. However, even in drug trials, new flexible methods have been developed that allow experiments to be stopped early or treatments to be discontinued early. Controversial at first, the methods are now the norm in many trials and are accepted by the US Food and Drug Administration, the regulatory body that must approve new drug applications.

One technique used maintains "alpha" at the traditional level of 5% (0.05) but allows it to be parceled out into multiple looks at the accumulating results. Rather than an experiment that must be run to a fixed sample size, the experiment can be phased. For example, the trial protocol might say that an initial assessment will be made with 100 patients in each group, and a (stringent) test of $p = 0.005$ is set as the determinant of significance. If the results do not prove significant, more patients are entered another assessment might be made at 300 patients in each group, say with a p-value threshold of 0.01. Additional assessments can be made, but each one counts against the overall allotment of $p = 0.05$. You are said to be "spending your alpha." This way, if a very strong treatment identifies itself early and beyond doubt, it can be brought to market immediately without waiting for the remainder of the trial to complete.

8.9.2 Business

Nonmedical businesses and other organizations have different motivations from drug companies when they run experiments. They are not seeking to assemble evidence in support of a new drug application but rather to gain information that helps them optimize and improve products, services, and products. Thus, they are not so constrained in seeking an escape from the straight-jacket approach of traditional hypothesis testing.

One new approach basically eliminates the strict dividing wall between experiments (to gather information) and deployment (cementing the results of experiments into operations). Experimentation becomes an ongoing iterative process, incorporating a feedback loop to modify the experiment. Operations then incorporate the new information on an ongoing basis. This approach is termed a bandit algorithm and is used in web testing where results are rapid and ongoing.

8.10 Bandit Algorithms

Bandit algorithms take their name from slot machines used in gambling, also termed one-armed bandits (since they are configured in such a way that they extract money from the gambler in a steady flow). If you imagine a slot machine with more than one arm, and each arm paid out at a different (but unknown) rate, you would have a multi-armed bandit, and that is the full name of this algorithm.

Your goal is to win as much money as possible and, more specifically, to identify and settle on the winning arm sooner rather than later. The challenge is that you don't know at what rate the arms pay out—you only know the results of pulling the arm. Suppose each "win" is for the same amount, no matter which arm. What differs is the probability of a win. Suppose further that you initially try each arm 100 times and get the following results:

- Arm A: 10 wins out of 50
- Arm B: 2 wins out of 50
- Arm C: 4 wins out of 50

One extreme approach is to say, "Looks like arm A is a winner—let's quit trying the other arms and stick with A." This takes full advantage of the information from the initial trial. If A is truly superior, we get the benefit of that early on. On the other hand, if B or C are truly better, we lose any opportunity to discover that.

Another extreme approach is to say, "This all looks to be within the realm of chance—let's keep pulling them all equally." This gives maximum opportunity for alternates to A to show themselves. However, in the process, we are deploying what seem to be inferior treatments. How long do we permit that?

Bandit algorithms take a hybrid approach: We start pulling A more often to take advantage of its apparent superiority, but we don't abandon B and C. We just pull them less often. If A continues to outperform, we continue to shift resources (pulls) away from B and C and pull A more often. If, on the other hand, C starts to do better and A starts to do worse, we can shift pulls from A back to C. If one of them turns out to be superior to A, and this was hidden in the initial trial due to chance, it now has a chance to emerge with further testing.

8.10.1 Web Testing

Now, think of applying this to web testing. Instead of three different slot machine arms, a merchant might have three different offers being tested on a website. Customers either click (a "win" for the merchant) or don't click. Initially, the offers are shown randomly and equally. If, however, one offer starts to outperform the others, it can be shown ("pulled") more often.

But what should be the parameters of the algorithm that modifies the pull rates? What "pull rates" should we change to, and when should we change?

Simulations similar to the resampling simulations we have seen, only more complex, can help answer these questions. Given assumptions about arm payouts, different switching rule algorithms can be simulated, and the probability of picking the best arm can be plotted as a function of time spent playing. The details of evaluating and implementing bandit algorithms are beyond the scope of this book, but an excellent short treatment, along with Python code to implement it, is in *Bandit Algorithms for Website Optimization* by John Myles White, O'Reilly, 2013.

8.11 Appendix: ANOVA, the Factor Diagram, and the *F*-Statistic

In this chapter, we looked at how to determine whether differences among groups are statistically significant. For measured data, we looked at ANOVA and the components of variance. Understanding the latter helps us quantify meaningful differences among groups, as opposed to random variation. For those interested, we now look in greater detail at how to unpack variance components.

8.11.1 Decomposition: The Factor Diagram

When we perform a decomposition, we find the means for each of our treatment groups and the grand average (or grand mean); see Table 8.12. We can take this information and put together a factor diagram as follows:

$$\underset{\text{Observations}}{\begin{pmatrix} 164 & 178 & 175 & 155 \\ 172 & 191 & 193 & 166 \\ 168 & 197 & 178 & 149 \\ 177 & 182 & 171 & 164 \\ 156 & 185 & 163 & 170 \\ 195 & 177 & 176 & 168 \end{pmatrix}} = \underset{\text{Grand average}}{\begin{pmatrix} 173.75 & 173.75 & 173.75 & 173.75 \\ 173.75 & 173.75 & 173.75 & 173.75 \\ 173.75 & 173.75 & 173.75 & 173.75 \\ 173.75 & 173.75 & 173.75 & 173.75 \\ 173.75 & 173.75 & 173.75 & 173.75 \\ 173.75 & 173.75 & 173.75 & 173.75 \end{pmatrix}}$$

$$+ \underset{\text{Treatment effect}}{\begin{pmatrix} -1.75 & 11.25 & 2.25 & -11.75 \\ -1.75 & 11.25 & 2.25 & -11.75 \\ -1.75 & 11.25 & 2.25 & -11.75 \\ -1.75 & 11.25 & 2.25 & -11.75 \\ -1.75 & 11.25 & 2.25 & -11.75 \\ -1.75 & 11.25 & 2.25 & -11.75 \end{pmatrix}} + \underset{\text{Residual error}}{\begin{pmatrix} -8.0 & -7.0 & -1.0 & -7.0 \\ 0.0 & 6.0 & 17.0 & 4.0 \\ -4.0 & 12.0 & 2.0 & -13.0 \\ 5.0 & -3.0 & -5.0 & 2.0 \\ -16.0 & 0.0 & -13.0 & 8.0 \\ 23.0 & -8.0 & 0.0 & 6.0 \end{pmatrix}}$$

Let's look at how the different components of the factor diagram were arrived at.

The Observation: In this part of our diagram, we are simply presenting the individual observations in our data set. Note that each column in the Observation portion of the diagram refers to a different group in our study (column 1 is Fat 1, column 2 is Fat 2, column 3 is Fat 3, and column 4 is Fat 4).

The Grand Average: In our example, the average of all the observations is 173.75. This is our baseline, or where the observation begins. Since the grand average is the same for all cases, we have just repeated the same value 24 times in the Grand Average portion of the diagram.

The Treatment Effect: For each group, we can find the group (treatment) mean and learn how much each group diverges from the grand average. This divergence is the same for all members of a group and is shown in the Treatment Effect portion of the diagram.

Note that if you now look across each row in the treatment effect portion of the diagram and add all the deviations (or treatment effects) together, they should sum or add to 0. In row 1, for example, we have $-1.75 + 11.25 + 2.25 + (-11.75) = 0$.

The Residual Error: Once we know the grand average and the treatment effect, what's left over is the residual error.

For example, remember that column 1 contains all doughnuts fried in Fat 1. The first case in that group had a value of 164. If we take 164 and subtract 172 from it (since 172 is the mean for Fat 1), we get −8.

8.11.2 Constructing the ANOVA Table

The decomposition table shows us how each case is made up of the different components: base average, group contribution, and residual error. Another table, called the ANOVA table, summarizes this information. Once we have decomposed the data, we can use the factor diagram to help us construct an ANOVA table and conduct the ANOVA. Conducting the ANOVA means to describe the variance components overall and assess whether group differences are significant. To construct our table, we need to determine:

- Degrees of freedom
- Sums of squares
- Mean squares
- An F statistic

Let's first examine how we would determine the appropriate degrees of freedom for each source of variability in our analysis.

The degrees of freedom (or df) are the number of observations that are theoretically free to vary once some parameter(s) describing the data is set.

For each section of our factor diagram, we can determine a corresponding value for degrees of freedom.

- For the **Grand Average** section, all values are, by definition, the same (since the grand average is the same for everyone). Once we know the grand average and place it in one cell of the Grand Average portion of the factor diagram, we know all other cells will have the same value. So, we only have 1 degree of freedom associated with the grand average.
- For the **Treatment Effect** section of the factor diagram, recall that the treatment effects must sum to 0 within each row. In this example, we have four treatment groups. Once we know three of our estimated treatment effects, the fourth must be that value that causes all treatment effects to sum to zero. There is also a formula we can use to determine the degrees of freedom due to treatment. It is the number of groups (denoted sometimes as "k") minus 1. In our example, we'd have $k - 1 = 4 - 1 = 3$.
- For the **Residual Error** section, we know that each column gives the deviations around the group mean. For each column, once we know all but one of the deviations, the last one is fixed (since the sum of the deviations must be 0). In this particular example, within each column, we'd have $6 - 1$ degrees of freedom,

or 5. Thus, the degrees of freedom associated with residual error would be 20 (or 5×4 since we have four groups). We can also use the formula $N - k$ to determine the degrees of freedom for residual error, where N is the total sample size (in this case, 24) and k is the number of groups (in this case, 4). We can see that $24 - 4 = 20$.

8.11.3 Inference Using the ANOVA Table

With our resampling procedure, we simply looked at the variance among the group means. In precomputer days, this was not possible. Just as with the *t*-test, a single standardized statistic that could be compared to tables was needed. Instead of the variance among the group means, this standardized statistic works from the ANOVA table and uses the *ratio* of two measures of variation, one for the treatment and the other for the residual. This allows us to use a single comparison table for all ANOVA problems.

We began our discussion about constructing the ANOVA table by illustrating ways in which we could determine degrees of freedom (df). To complete our ANOVA table, we also need to obtain the sum of squared deviations, or sum of squares for short, for the treatment (or independent variable[3]) groups and for the residuals. Actually, we will deal with averages more than sums, i.e. *mean squares or MS*. From these, we can calculate the F-statistic and, from that, a p-value.

> **Why SQUARED deviations?**
> We square deviations before summing them to render them all positive; if we simply added deviation values together in the factor diagram without squaring them, the sum would always end up being zero and all sense of variation would be lost. (We could take absolute values of the deviations, but squared values are easier to deal with mathematically in calculations.)

To find the sum of squares for our treatment effects, we would square each cell in that section of the diagram and add all the squared values together. We do the same for the residual error section.[4] If you do this, you should get:

- Sum of Squares (Treatment Effect) = 1636.5
- Sum of Squares (Residual Error) = 2018

3 Also called the *predictor* variable.
4 Some ANOVA tables also reference the sum of squares, or SS, for the grand average, which is the sum of all the (identical) values in that section of the diagram squared.

We obtain the mean squares for treatment by taking the sum of squares for the treatment effect (or 1636.5) and dividing it by the degrees of freedom for the treatment effect (or 3). This gives us 1636.5/3 = 545.5. We can also find the mean square for our residual error by taking the sum of squares for the residual error (or 2018) and dividing it by the degrees of freedom for the residual error (or 20). This gives us 2018/20 = 100.9.

We get our F-**statistic** by dividing the mean square for our treatment effect by the mean square for our residual error. In this example, we get $F = 545.5/100.9 = 5.41$.

The formula for the F-statistic is:

$$F = \frac{\text{MS}_{\text{Treatment}}}{\text{MS}_{\text{Residual}}}$$

In Figure 8.6, you will see the ANOVA table that was created using `statsmodels`. The larger the F-value, the greater the contribution of the treatment variable to overall variability and the less likely it is that chance is responsible for the differences among groups. The p-value is computed for us automatically in `statsmodels`. It is based on a comparison of the observed F-value to a stored table of F-values that reflect what we would obtain by resampling.

8.11.4 The *F*-Distribution

We saw earlier that W. S. Gossett devised a formula for the t-distribution that approximates the distribution of a sample mean and the difference between two sample means. There is an analogous distribution for the ratio of treatment and error variances.

To determine if we have a significant F statistic (and to estimate our p-value), we compare our calculated F statistic to a standard distribution of variance ratios. To do this, we need a family of standard distributions of these ratios. This family is called the "F distribution," after Sir Ronald Fisher, who originally proposed the formal ANOVA. Each distribution in this family is characterized by two separate values for degrees of freedom (df), where k is the number of groups:

1) df for the numerator (or treatment effect) is $= k - 1$.

```
                df   sum_sq  mean_sq         F    PR(>F)
C(variable)    3.0   1636.5    545.5  5.406343  0.006876
Residual      20.0   2018.0    100.9       NaN       NaN
```

Figure 8.6 Resulting ANOVA table from `statsmodels`.

2) df for the denominator (or residual error) is = $k(n - 1)$, (where n = is the number of values in each sample. Where there are an unequal number of cases per sample, the more general $(N - k)$, where N is the total sample size across all groups, is used.

As noted earlier, based on our decomposition of the data, the degrees of freedom for the numerator are 3 and the degrees of freedom for the denominator are 20. We write this as $F_{(3,20)}$, and this refers to the F distribution for four groups of six values each.

You can compare the observed F value (5.406) to a critical F value by looking it up in a web calculator, or you can use Python to determine the area to the right of the observed F statistic using the `scipy` functions *stats.f.cdf* (cumulative distribution function) or *stats.f.sf* (survival function). The two methods are equivalent.

```
from scipy import stats
x = 5.406
print(f"p-value {1 - stats.f.cdf(x, 3, 20):.4f}")
print(f"p-value {stats.f.sf(x, 3, 20):.4f}")
```

Output

```
p-value 0.0069
p-value 0.0069
```

Figure 8.7 visualizes the probability density and the cumulative distribution density of $F_{(3,20)}$. It shows that the p-value is very small. This matches closely with the p-value we saw earlier in our `statsmodels` output and in the results from using a resampling procedure.

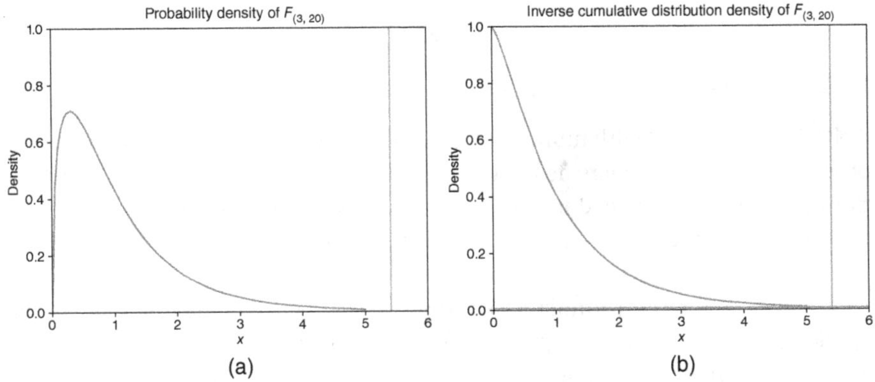

Figure 8.7 Probability density of $F_{(3,20)}$ (a) and its inverse cumulative density (b). The vertical line shows the F-statistic obtained for the fat dataset. The horizontal line shows the corresponding p-value.

Note that the statistic that we used in our resampling procedure was the variance among group means. We could also have used the F statistic itself as our resampling statistic, calculating it during each resample and comparing the resampled distribution to the observed value. Fisher, like Gossett, simply found a mathematical approximation to the resampled distribution and reduced it to tables of values (these F tables, along with z tables, t tables, and more, are typically found at the back of most statistics texts).

8.11.5 Different Sized Groups

So far, we have only dealt with situations where the groups are all equally sized. What about situations where the group sizes are different?

8.11.5.1 Resampling Method

When group sizes differ, only step two in the resampling method changes, and then only slightly. Instead of shuffling the box and drawing four groups of six values each, as before in step 2, we now shuffle the box and draw four groups with the same number of values as in each original group. So, as an example, suppose one group has three values, a second has four values, a third has six values, and a fourth group has seven values. For this case, we shuffle the box and draw three values for the first group, four for the second, and so forth.

8.11.5.2 Formula Method

In contrast to the resampling method, the formula method described above only works when the groups are all the same size. If the groups are not the same size, the computations are messier and harder to interpret, though some statistical software handles this "under the hood."

8.11.6 Caveats and Assumptions

ANOVA may be used with multigroup controlled experiments or with multigroup observational studies where no treatment is involved. For experiments, treatments must be randomly assigned to subjects. For observational studies, sample groups must be independently and randomly selected from a much larger population of interest.

In addition, for the formula-based approach, which includes most software applications, we assume our groups are similar in terms of variability (e.g. we assume the largest group standard deviation is no bigger than twice the smallest group standard deviation), and we assume the outcome variable is Normally distributed. These requirements can be checked by comparing standard deviations among the groups, by graphical analysis, or by examining residuals.

8.12 More than One Factor or Variable—From ANOVA to Statistical Models

Running an experiment is expensive, and it is often cost-effective to study more than one thing at a time. We can add another variable to ANOVA, which is then termed two-way ANOVA. Two-way ANOVA is more complex than standard one-way ANOVA and is beyond the scope of this book. Instead, we turn to statistical and machine learning methods that model the relationship between input predictor variables (factors) and a target (outcome) variable. The emphasis becomes not on *whether* a variable has an effect but on *how* that variable affects an outcome. After we look at correlation in the next chapter, the rest of this book will introduce you to those modeling methods.

8.13 Python: Contingency Tables and Chi-square Test

8.13.1 Example: Marriage Therapy

We already learned how to create two-way tables using Python in Section 6.7 using the *pd.crosstab* function. Let us recap this with the *marriage-therapy.csv*.

```
import pandas as pd
marriage_therapy = pd.read_csv("marriage-therapy.csv")
contingency_table = pd.crosstab(marriage_therapy["Therapy"],
marriage_therapy["Outcome"])
contingency_table
```

Output

```
Outcome      Distress Married  Divorced  Happily Married
Therapy
Behavioral                  3        11               15
Insight                     5         1               24
```

In this chapter, we learned how to use resampling or the chi-square test to decide whether the therapy option influences the outcome or not. For this, we need to be able to calculate the expected outcomes if the treatment option does not influence the outcome. We write a function that returns the expected table given the observed table. Here is the first version of this function:

```
def expected_table(observed_table):
    total = observed_table.sum().sum() ①
    row_prob = observed_table.sum(axis=1) / total ②
    col_prob = observed_table.sum(axis=0) / total
    expected = pd.DataFrame(index=observed_table.index,
                            columns=observed_table.columns) ③
```

```
    for row in observed_table.index: ④
        for col in observed_table.columns:
            expected.loc[row, col] = row_prob[row] * col_prob[col] * total
    return expected
```

① The total number of observations is the sum of all entries in the table. For a pandas DataFrame, we need to call the *sum* method twice, once for the rows and once for the columns.
② We calculate the row and column probabilities by summing over the rows and columns and dividing by the total number of observations.
③ We create an empty pandas DataFrame with the same row and column labels as the observed table.
④ We iterate over all rows and columns and calculate the expected value for each cell by multiplying the row and column probabilities (assumption of independence) and the total number of observations.

The implementation can be made more efficient by realizing that the expected value is the *outer product* of the probabilities multiplied by the total number of observations. Here is a version that uses the numpy function *np.outer* to calculate the outer product.

```
import numpy as np
def expected_table(observed_table):
    total = observed_table.sum().sum()
    row_sums = observed_table.sum(axis=1) ①
    col_sums = observed_table.sum(axis=0)
    expected = np.outer(row_sums, col_sums) / total ③
    return pd.DataFrame(expected,
             index=observed_table.index, columns=observed_table.columns) ④

expected_table(contingency_table).round(2)
```

① Instead of calculating the row and column probabilities, we calculate the row and column sums. This reduces the number of calculations.
② The *np.outer* function is called with the row and column sums. The outer product is then divided by the number of observations to get the expected values.
③ We create a pandas DataFrame from the numpy array and return it.

Output

```
Outcome      Distress Married  Divorced  Happily Married
Therapy
Behavioral               3.93       5.9            19.17
Insight                  4.07       6.1            19.83
```

With the function in place, we can now implement the resampling experiment from Section 8.1.3. The resampling procedure is:

```
import random
random.seed(123)
```

```
def calculate_difference(contingency_table): ①
    expected = expected_table(contingency_table)
    difference = contingency_table - expected
    return difference.abs().sum().sum()

def resample(therapy_box, outcome_box): ②
    random.shuffle(outcome_box)
    contingency_table = pd.crosstab(therapy_box, outcome_box)
    return calculate_difference(contingency_table)

outcome_box = (["Happily Married"]*39 + ["Distress Married"]*8 +
               ["Divorced"]*12)
therapy_box = ["Behavioral"]*29 + ["Insight"]*30

resamples = np.array([resample(therapy_box, outcome_box)
                      for _ in range(10000)]) ③

observed = calculate_difference(contingency_table)

print(f"Observed difference: {observed:.2f}")
print("Resamples above observed difference: "
      f"{np.sum(resamples >= observed)}")
print("p-value:", np.mean(np.array(resamples) >= observed))
```

① The `calculate_difference` function calculates the difference between a contingency table and the expected table under the assumption of independence.

② The `resample` function implements the resampling procedure. It takes the therapy and outcome boxes as arguments. It shuffles the outcome box and creates a new contingency table. It then calculates the expected table and the difference between the observed and expected tables. The sum of the absolute values of the differences is returned.

③ We call the *resample* function 10,000 times and store the results in a list.

Output

```
Observed difference: 20.41
Resamples above observed difference: 68
p-value: 0.0068
```

The low *p*-value indicates that the observed difference is very unlikely under the assumption of independence. Figure 8.8 shows this visually.

```
ax = pd.Series(resamples).plot.hist(bins=20)
ax.axvline(observed, color="red")
ax.set_xlabel("Sum of absolute differences to expected table")
ax.set_ylabel("Frequency")
plt.show()
```

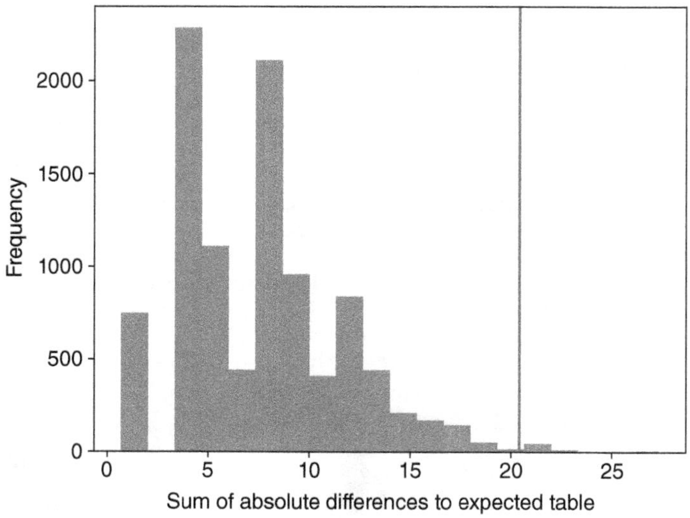

Figure 8.8 Resampling procedure for the marriage therapy data. The observed difference is shown as a vertical line.

8.13.2 Example: Imanishi-Kari Data

As a second example, we use resampling to analyze the Imanishi-Kari data. Here is the Python code that implements the resampling procedure from Section 8.2.

```
from collections import Counter
random.seed(123)

box = list(range(10)) ①

differences = []
for _ in range(10_000):
    random.shuffle(box)
    resample = random.choices(box, k=315)
    counts = Counter(resample) ②
    difference = sum(abs(counts[i] - 31.5) for i in range(10)) ③
    differences.append(difference)
differences = np.array(differences)
above_216 = sum(differences >= 216)
p_value = above_216 / len(differences)

print("Number of resamples with sum of absolute deviations >= 216: "
      f"{above_216}")
print(f"p-value = {p_value:.4f}")
```

① We create a list with the digits from 0 to 9.

② We shuffle the box and draw 315 digits from it with replacement. We then count the number of occurrences of each digit type using the `Counter` class.
③ We calculate the sum of the absolute differences between the observed counts and the expected counts (31.5).

Output
```
Number of resamples with sum of absolute deviations >= 216: 0
p-value = 0.0000
```

None of our resampled differences is larger than 216, and the *p*-value is 0.0000. The observed distribution of digits is highly unusual.

8.14 Python: ANOVA

8.14.1 Visual Comparison of Groups

In this chapter, we learned about ANOVA. It is possible to use the `statsmodels` package to perform ANOVA, but let us first see how we can perform an ANOVA with Python using the fat absorption data as an example. We start by loading the data and creating a variety of visualizations.

```
fat_absorption = pd.DataFrame([
    [1, 164, 178, 175, 155], ①
    [2, 172, 191, 193, 166],
    [3, 168, 197, 178, 149],
    [4, 177, 182, 171, 164],
    [5, 156, 185, 163, 170],
    [6, 195, 177, 176, 168],
], columns=["Replication", "Fat 1", "Fat 2", "Fat 3", "Fat 4"]) ②
```

① The pandas `DataFrame` is created from a list where each list element is the result of one replication.
② The `columns` argument defines the column labels.

Next, we visualize the data. Figure 8.9a shows the data in the form of a dot plot.

```
fats = ["Fat 1", "Fat 2", "Fat 3", "Fat 4"]

# convert from wide to long format for plotting
fat_long = fat_absorption.melt(value_vars=fats, var_name="group")①

fig, ax = plt.subplots(figsize=(5, 3))
fat_long.plot.scatter(x="group", y="value", ax=ax)
ax.set_xlabel("Fat type")
ax.set_ylabel("Grams of fat absorbed")
ax.set_xlim(-0.5, 3.5)
plt.show()
```

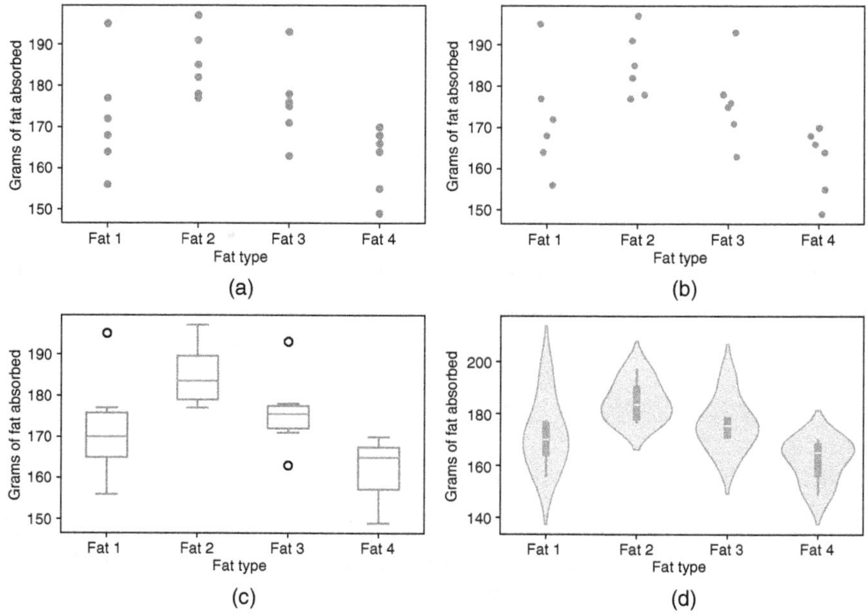

Figure 8.9 Visualizations for the doughnut experiment. (a) Dot plots, (b) dot plot with jitter, (c) boxplots, and (d) violin plot.

① We use the *melt* method to convert the `DataFrame` from wide to long format. The resulting dataframe has two columns. The `group` column contains the column names, and the `value` column contains the values of the original `DataFrame`. To convert from long to wide format, use the *pivot* method.

This type of visualization works only for a small number of data points so that the probability of overlapping points is low. We already learned how we can use transparency in this case by setting the argument `alpha` of the *plot* function to a value less than 1. An alternative is to add random displacement to the points. This is known as adding jitter. The *stripplot* function from the `seaborn` package has this functionality. We enable it by setting `jitter=True`. The resulting graph is Figure 8.9b.

```
import seaborn as sns
fig, ax = plt.subplots(figsize=(5, 3))
sns.stripplot(data=fat_long, x="group", y="value", color="C0",
jitter=True, ax=ax)
ax.set_xlabel("Fat type")
ax.set_ylabel("Grams of fat absorbed")
ax.set_xlim(-0.5, 3.5)
plt.show()
```

Boxplots are another good way to visualize the distribution of a group for different groups (see Figure 8.9c)

```
fig, ax = plt.subplots(figsize=(5, 3))
fat_absorption[fats].plot.box(ax=ax) ①
ax.set_xlabel("Fat type")
ax.set_ylabel("Grams of fat absorbed")
plt.show()
```

① The *plot.box* method of the `DataFrame` creates the boxplot for each column. This is the reason why we use the `fats` list to select the columns. We could have also created the boxplot using the `fat_long` data: `fat_long .plot.box("group")`.

Boxplots can hide the actual distribution of the data. If you want to see those, you can use a violin plot (see Figure 8.9d). The `seaborn` package has the *violinplot* function to create these.

```
import seaborn as sns
fig, ax = plt.subplots(figsize=(5, 3))
sns.violinplot(data=fat_long, x="group", y="value", color="lightgrey", ax=ax)
ax.set_xlabel("Fat type")
ax.set_ylabel("Grams of fat absorbed")
ax.set_xlim(-0.5, 3.5)
plt.show()
```

8.14.2 ANOVA Using Resampling Test

Now that we have visualized our data, we can perform the resampling experiment from Section 8.3. We start by writing a function that calculates the *deviation* of the group means from the total mean. We use the data in the long format.

```
def calculate_deviation(data, group, values): ①
    total_mean = data[values].mean() ②
    group_means = data.groupby(group)[values].mean() ③
    k = len(group_means)
    return sum((group_means - total_mean)**2) / (k - 1) ④

deviation = calculate_deviation(fat_long, "group", "value")
print(f"Deviation: {deviation:.2f}")
```

① The `calculate_deviation` function takes three arguments: the `DataFrame`, the name of the column that contains the group labels, and the name of the column that contains the values.
② This calculates the total mean of all values
③ The *groupby* method splits the dataframe into subsets using the unique values in the `data`. We then reduce it to the `values` and calculate the *mean()* of the values in each subset. At the end, we have a `Series` with the means of each group.
④ Finally, we calculate the variance among the group means and return it.

Output
```
Deviation: 90.92
```

We can now implement the resampling procedure.

```
rng = np.random.default_rng(seed=4321) ①
observed = calculate_deviation(fat_long, "group", "value")
resamples = []
box = fat_long.copy() ②
for _ in range(10_000):
    shuffled = box["group"].sample(frac=1, random_state=rng) ③
    box["group"] = shuffled.to_numpy() ④
    resamples.append(calculate_deviation(box, "group", "value"))
resamples = np.array(resamples)
p_value = np.mean(resamples >= observed)

print(f"Observed deviation:{observed:.2f}")
print(f"Resamples above observed deviation: {np.sum(resamples >= observed)}")
print(f"p-value: {p_value:.4f}")
```

① Set a random seed for reproducibility.
② We create a copy of the original dataframe, to avoid changing the original dataframe.
③ This statement shuffles the content of the `group` column. The result is a `pandas Series`.
④ Even though the order of the groups changed, `pandas` keeps the index of the original dataframe and would assign them back into the column in the original order. Changing the `Series` to a numpy array using the function *to_numpy*, we remove the index and are now able to assign the shuffled group back to the `group` column.

Output
```
Observed deviation:90.92
Resamples above observed deviation: 84
p-value: 0.0084
```

The p-value of 0.0084 is very low, which means that the observed difference between the fats is very unlikely under the assumption that the means of the groups are equal. The distribution of the resampled variances, shown in Figure 8.10 confirms this interpretation.

```
ax = pd.Series(resamples).plot.hist(bins=30)
ax.set_xlabel("Variance")
ax.axvline(observed, color="black")
plt.show()
```

8.14.3 ANOVA Using the *F*-Statistic

ANOVA is such an important statistical test that it is implemented in many statistical packages. We can use the `statsmodels` package to perform an ANOVA. Here is the code for the doughnut data.

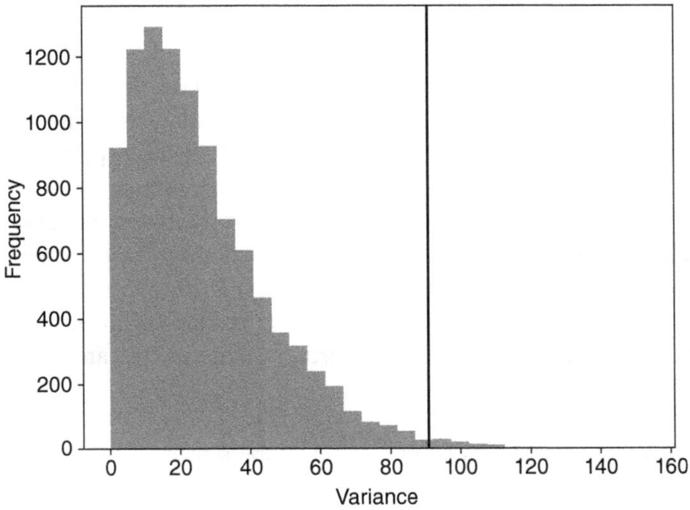

Figure 8.10 Frequency histogram of resampled variances from doughnut data problem.

```
import statsmodels.api as sm ①
from statsmodels.formula.api import ols

model = ols("value ~ C(group)", data=fat_long).fit() ②
table = sm.stats.anova_lm(model) ③
print(table)
```

① The `statsmodels` package is commonly imported as `sm`.

② We use the *ols* function to create a model. The formula `"value ~ C(group)"` specifies that we want to use the `value` column as the dependent variable and the `group` column as the independent variable. The `C` function is used to indicate that `group` is a categorical variable. The model is fitted using the `fat_long` data.

③ The fitted `model` is passed to the *sm.stats.anova_lm* function. The result is a `pandas DataFrame` with the ANOVA table.

Output

```
                df  sum_sq  mean_sq         F    PR(>F)
group          3.0  1636.5    545.5  5.406343  0.006876
Residual      20.0  2018.0    100.9       NaN       NaN
```

The F-statistic is 5.41. The p-value calculated from the F-statistic is 0.0069, which is very close to the p-value we calculated using the resampling test.

Exercises

8.1 A nursery wholesaler breeds flowers, and a botanist believes that a certain cross will produce plants with either white, red, or pink flowers in the following proportions: 60% white, 30% red and 10% pink. A given plant has only one color. The botanist is experimenting with a treatment that may alter the above proportions. Further development of such a treatment would be useful in helping match flower color to customer demand if the treatment does, in fact, alter the proportions of flowers. In a test of 100 plants, 65 produced white flowers, 24 produced red flowers, and 11 produced pink. Is there evidence that the treatment alters color proportions? State and test an appropriate hypothesis.

8.2 A social media site that allows community members to rate restaurants wants to ensure that the reviews are genuine and not spurious. For each restaurant, every four days, it collects reviews and counts the "favorable" reviews. Per a consultant's recommendation, for each restaurant, it performs a hypothesis test on the four-day data to determine whether the proportion "favorable" is outside the range of expected chance variation, given a historic favorability benchmark for the restaurant. Is there a multiple testing issue in this procedure? If so, what effect would that have on interpreting a given hypothesis test?

8.3 A retailer of bike racks for vehicles conducts an A/B test for how to present one of its products online. A randomizer routes visitors to one of the two treatments. The outcomes of interest are "bounce" (exit the page with no further exploration), "click" (click on a link to further explore the product, but no purchase), or "purchase" (continue through exploration to purchase before the expiration of cookies that track a user). Consider the data in Table 8.14 on two alternate treatments.

Table 8.14 User responses to two online product treatments.

	Bounce	Click	Purchase
Treatment A	312	69	7
Treatment B	279	87	2

Is there a statistically significant difference between the two online product treatments?

8.4 Web page load time depends on a number of variables, but internet service providers (ISPs) have control only over factors at the server end. An ISP is considering whether to invest staff time in working on different configurations and deploying more successful ones. The first step is an experiment in web page load times with three different server configurations, A, B, and C. Different web pages are tested at a remote location with the three configurations, each page being tested under equal conditions for A, B, and C. A limited number of pages are tested, since each one requires a build on three different servers. The web page load times are shown in Table 8.15 and are available in the *server-configurations.csv* dataset.

Table 8.15 Web page load times in seconds for 3 different server configurations.

A	B	C
1.049	1.979	0.948
1.029	1.992	0.856
1.349	2.323	1.004
1.171	2.056	0.386
2.071	2.308	0.308
0.366	3.558	1.589
1.201	2.206	0.274
1.434	2.291	0.767
0.836	1.962	0.978
3.201	1.112	0.371
0.415	2.368	0.298
1.934	1.178	0.445

a) Calculate the average load time at the three different server configurations.
b) Visualize the data to compare the three configurations.
c) Test the hypothesis that the three servers do not differ with respect to load time using resampling.
d) Test the hypothesis that the three servers do not differ with respect to load time using the ANOVA implementation of the `statsmodels` package.

9

Correlation

In this chapter, we look at relationships between variables. After completing this chapter, you should be able to:

- Define what we mean by correlation
- Determine the statistical significance of apparent correlation by resampling
- Calculate the correlation coefficient
- Assess whether correlation suggests causation

Think back to the "no-fault" study of errors in hospitals. In an experiment, we found that introducing a no-fault reporting system reduced the number of serious errors. We found there was a relationship between one variable—a type of reporting system—and another variable—a reduction in errors.

"Type of reporting system" is a binary variable. It has just two values: "regular" and "no-fault." Often, you may want to determine whether there is a relationship involving the *amount* of something, not just whether it is "on" or "off."

For example, is there a relationship between employee training and productivity? Training is expensive, and organizations need to know not simply *whether* training helps but also *how much* it helps.

Definition: Correlation Correlation is an association between the magnitude of one variable and that of a second—e.g. as x_1 increases, x_2 also increases (positive correlation). Or as x_1 increases, x_2 decreases (negative correlation). Different statistics are used to measure correlation; we will develop and define two in this chapter.

9.1 Example: Delta Wire

In the 1990s, Delta Wire, a small wire manufacturing company in Clarksdale, Mississippi, started a basic skills training program. Over time, as wastage declined,

Statistics for Data Science and Analytics, First Edition. Peter C. Bruce, Peter Gedeck, and Janet Dobbins.

Companion website: www.wiley.com/go/Wiley_Statistics_for_Data

Table 9.1 Training and productivity at Delta Wire.

Total Training Hours Cumulative	Productivity Pounds per Week
0	70,000
100	70,350
250	70,500
375	72,600
525	74,000
750	76,500
875	77,000
1100	77,400
1300	77,900
1450	77,200
1660	78,900
1900	81,000
2300	82,500
2600	84,000
2850	86,500
3150	87,000
3500	88,600
4000	90,000

production at the facility improved from 70,000 pounds of wire per week to 90,000 pounds per week. Table 9.1 shows some detailed data[1] from the program.

Clearly, more training led to higher productivity throughout the period. We can get a better idea of the relationship from the scatterplot in Figure 9.1. Henceforth, we will deal with productivity in terms of thousands of pounds per week for easier display.

Figure 9.1 presents a clear picture: more training is associated with higher productivity. No further analysis is needed to reach this conclusion. We can say that the correlation is positive, and the relationship is more or less linear.

---

1 The Delta Wire case was reported in Bergman, Terri, "TRAINING: The Case for Increased Investment," *Employment Relations Today*, Winter 1994/95, pp. 381–391. Detailed data are from an adaptation by Ken Black, *Business Statistics*, Wiley, Hoboken, NJ, 2008, p. 589.

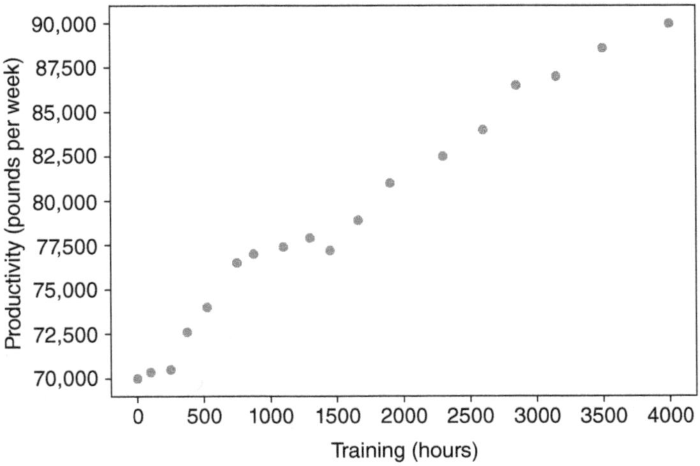

Figure 9.1 Delta Wire productivity vs. training.

9.2 Example: Cotton Dust and Lung Disease

Figure 9.2, which illustrates the respiratory function of workers in cotton factories in India, depicts a relationship that is less clear.[2] The *y*-axis shows a worker's Peak Expiration Flow Rate (PEFR), and higher is better. The *x*-axis shows years of exposure to cotton dust in a cotton factory.

Is PEFR related to Exposure? It's hard to tell, just based on the picture. We need:

1) A way to measure the correlation between two numeric variables, and
2) A way to determine whether the correlation we measure is real or might just be a chance association.

We will tackle these questions in the next section. But before we proceed:

Can you spot an unusual feature of the exposure variable? What might account for it?

2 Diagram is based on data published in "To Know the Prevalence of Byssinosis in Cotton Mill Workers & to know Changes in Lung Function in Patients of Byssinosis," *Indian Journal of Physiotherapy and Occupational Therapy*, Authors: Sarang Bobhate, Rakhi Darne, Rupali Bodhankar, Shilpa Hatewar, Vol. 1, No. 4 (2007-10–2007-12); available as of January 2024 at https://ijpot.com/scripts/Oct-Dec\LY1\textbackslash%202007\LY1\textbackslash %20issue\LY1\textbackslash%20of\LY1\textbackslash%20IJPOT.pdf.

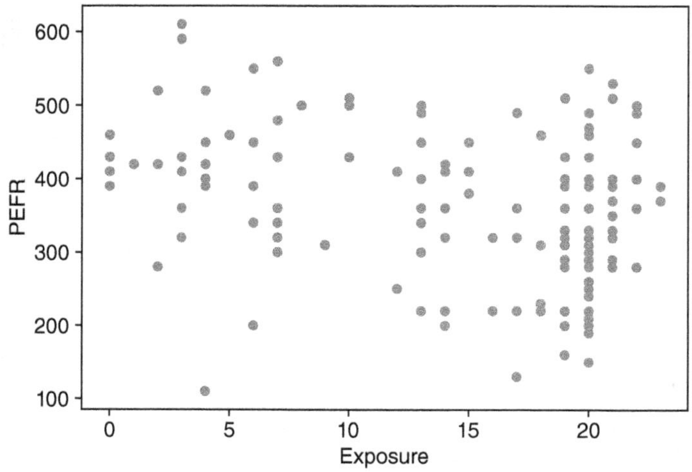

Figure 9.2 Pulmonary capacity (PEFR) and exposure to cotton dust (years).

9.3 The Vector Product Sum Test

Vectors **x** and **y** are correlated if high values of **x** correspond to high values of **y**. One way to measure the correlation between **x** and **y** is to multiply them together as vectors and then sum them. That sum is the *vector product sum*, and it is greatest when the rank order of **x** matches that of **y**.

> **Vector and Matrix Notation**
> A *vector* is a list of numbers—a single column or row. It is denoted by a lower-case bold letter, as with **x** and **y** above. A *matrix* is a two-dimensional array of numbers with both rows and columns. It is denoted by an uppercase bold letter, e.g. **X**. We have not yet dealt with matrices.

Definition: Rank Order Rank Order is a list of the rank positions of a list of values. For example, consider the list {7 9 6 3}. The rank order of this list is {3 4 2 1}.

> **Try It Yourself**
> Let **x** be [1 2 3] and **y** be [2 3 4]. Note that they are in the same rank order—lowest to highest. When multiplied together, the vector product sum is 20.

x	y	Product
1	2	2
2	3	6
3	4	12

$$\text{Vector product sum} = 2 + 6 + 12 = 20$$

Let's try a different arrangement:

x	y	Product
2	2	4
1	3	3
3	4	12

$$\text{Vector product sum} = 4 + 3 + 12 = 19$$

Try rearranging the **x** vector in various other ways and recalculating the vector product sum. You will see that it is never as high as it is when the two vectors are in the same rank order.

Note: In the same way, the vector product sum is smallest when the two variables' rank orders are exact opposites–i.e. when they are perfectly negatively correlated.

Definition: Perfect Correlation Perfect positive correlation is when the rank order of one variable exactly matches the rank order of the second variable—highest value goes with highest value, second-highest with second-highest, etc. Perfect negative correlation is when the rank order of one variable is the exact opposite of the second variable—highest value goes with lowest, second-highest goes with second-lowest, etc.

Your "Try it Yourself" results, with the arrangements of [1 2 3] and [2 3 4], suggest a way of determining whether an apparent correlation between two variables might have happened by chance.

1) Write down the values for one of the variables on a set of cards.
2) Write down the values for the other variable on a second set of cards.
3) Array the two sets of cards next to each other, one column for variable one and the other column for variable two. Make sure that cards for the same case are

adjacent to one another. For example, in the lung disease case, the card for a 110 PERF score must be next to a card for four-year exposure. Multiply the two variables, and sum those values to calculate the vector product sum.

4) Shuffle one set of cards, repeat the multiplication, and record the vector product sum.
5) Repeat the above step 1000 times.
6) Find out how often the shuffled sum is equal to or greater than the observed sum.

The result is the *p*-value. Given the null model of no correlation, the *p*-value is the probability that a vector product sum as large or larger than the observed value might occur by chance.

9.3.1 Example: Baseball Payroll

Is the total salary payroll of a U.S. Major League Baseball team correlated with performance? More specifically, do teams with higher payrolls win more games (Figure 9.3)?

Table 9.2 shows actual data for several teams, along with the products and the vector product sum.

9.3.1.1 Resampling Procedure

1) Multiply the payroll vector by the wins vector. The sum of products is 668,620.
2) Shuffle one vector—we will shuffle the wins vector—and re-multiply. Record the shuffled sum.
3) Repeat step 2 many times (say 1000).
4) How often do we get a sum of products >= 668,620?

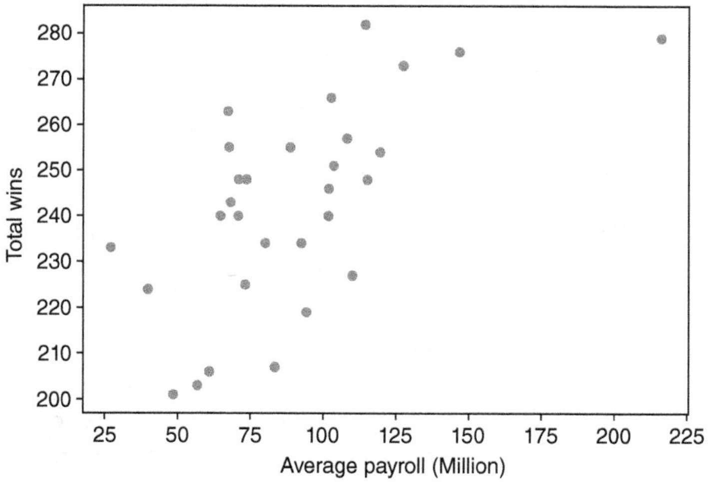

Figure 9.3 Baseball payroll vs. total wins, 2006–2008.

Table 9.2 Excerpt of baseball payroll and total wins.

Team Name	Average Payroll (Million $)	Total Wins	Product
Yankees	216.1	279	60,292
Red Sox	146.66	276	40,478
Mets	127.4	273	34,780
.	.	.	.
.	.	.	.
.	.	.	.
Rays	39.76	224	8906
Marlins	27.07	233	6307
		Vector product sum = 668,620	

From the histogram in Figure 9.4, we can see that a product sum of 668,620 rarely, if ever, occurs. So, we conclude that the degree of correlation between payroll and wins is not just a coincidence.

Try It Yourself

An online dating site seeks to learn more about what it can do to encourage successful outcomes in the relationships formed by its customers. It collects various data, including satisfaction surveys six months after a customer first signs up and also how much time the customer has spent on its dating site. Satisfaction data is recorded as an integer between 1 (low satisfaction) and 10, and time-on-site is recorded in minutes. Here are hypothetical results:

Time Spent	Satisfaction
10.1	2
67.3	7
34.0	2
2.9	1
126.3	9
39.0	8
4.6	1
211.3	6

Calculate the vector product sum and use a resampling procedure to test whether there is a correlation between time spent and satisfaction.

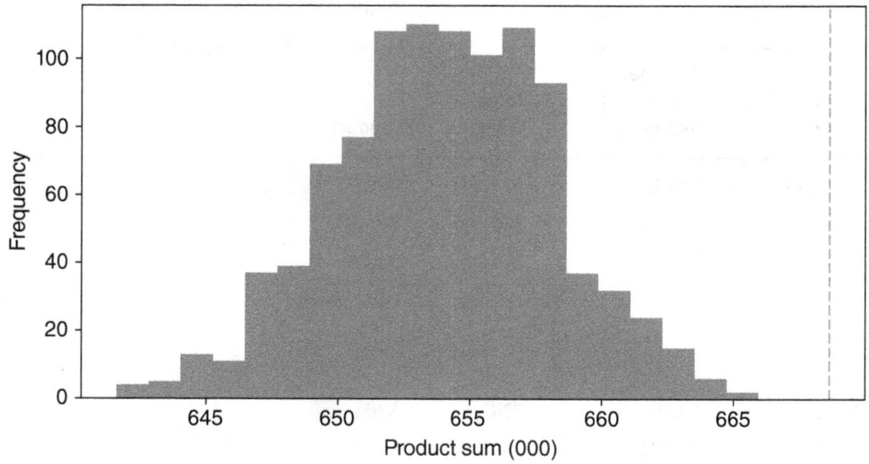

Figure 9.4 Baseball histogram of shuffled vector product sums (000).

9.4 Correlation Coefficient

The vector product sum can be used to test the statistical significance of the correlation between two variables, but the sum itself is not that meaningful.

Consider the vector product sum for the three examples above.

- Baseball: 688,620
- Lung disease: 603,670
- Worker training: 2,397,534.

These numbers cannot be compared to one another, nor can they be used to measure the strength of correlation.

Instead, statisticians use a standardized version of the vector product sum called the correlation coefficient. As we saw previously, standardization involves subtraction of the mean and division by the standard deviations.

Definition: Correlation Coefficient Let x and y denote the two variables. Consider a sample of size n of paired data (X_i, Y_i). Then, the sample correlation coefficient, denoted as r or the Greek letter ρ (rho), is given by

$$r = \frac{\sum_{i=1}^{n} \left(X_i - \overline{X}\right)\left(Y_i - \overline{Y}\right)}{(n-1)\, s_x s_y}$$

where $(\overline{X}, \overline{Y})$ are the sample means and (s_x, s_y) are the sample standard deviations for x and y.

As a result of standardization, the correlation coefficient always falls between −1, which is a perfect negative correlation, and +1, which is a perfect positive correlation. A value of zero indicates no correlation.

Note: The correlation coefficient defined above is called the "Pearson product moment correlation coefficient" or simply the "Pearson correlation coefficient." It is named after the great English statistician Karl Pearson (1857–1936). There are other correlation coefficients, but this one is the most widely used.

The correlation coefficients for the three examples cited above follow below. The $x - y$ scatterplots are repeated for reference.

1) Baseball: $r = 0.63$ (Figure 9.5)
2) Lung Disease: $r = -0.28$ (Figure 9.6)
3) Worker Training: $r = 0.99$ (Figure 9.7)

For the worker training example, the 0.99 value for the correlation coefficient indicates a very strong positive correlation between training time and productivity.

For the lung disease example, the −0.28 value for the correlation coefficient indicates modest negative correlation between exposure time and lung function (more exposure leads to lesser lung function).

For the baseball example, the 0.64 value for the correlation coefficient indicates strong positive correlation between payroll and won-loss record.

9.4.1 Inference for the Correlation Coefficient—Resampling

Could the correlation simply be the result of random chance?

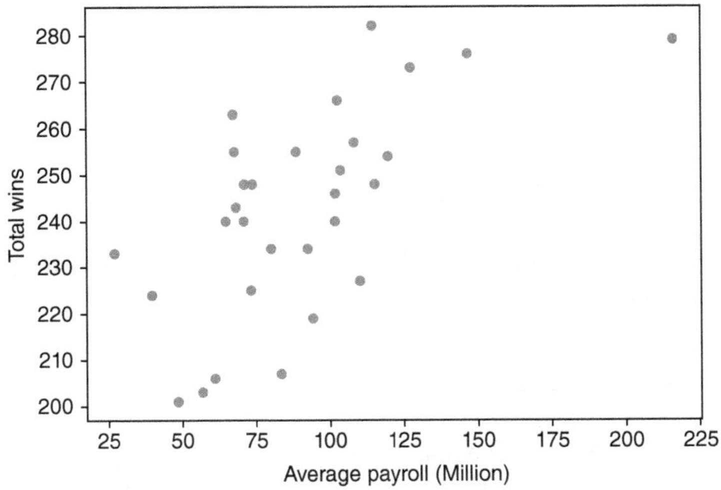

Figure 9.5 Baseball payroll vs. total wins, 2006–2008; $r = 0.63$.

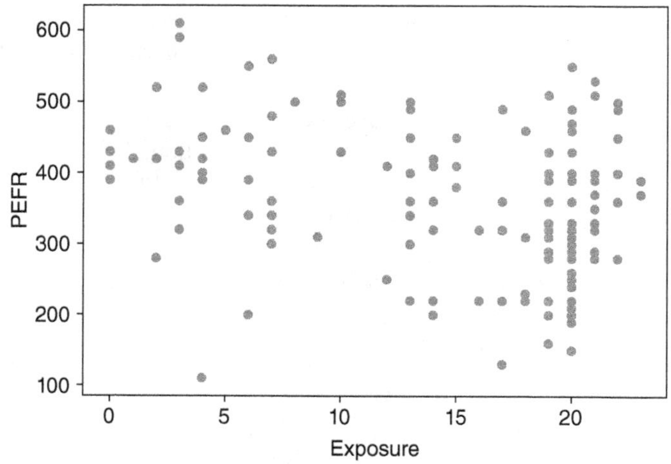

Figure 9.6 Pulmonary capacity (PEFR) and exposure to cotton dust (years); $r = -0.28$.

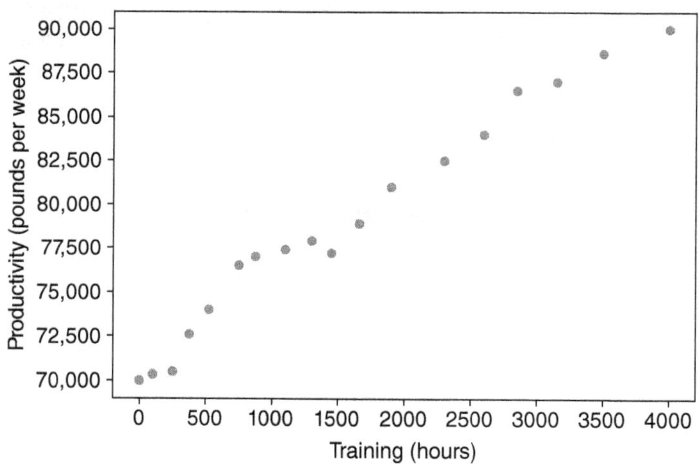

Figure 9.7 Delta Wire productivity vs. training; $r = 0.99$.

Significance testing for correlation has already been illustrated for the vector product sum method. We can use the same procedure with the correlation coefficient.

9.4.1.1 Hypothesis Test—Resampling

1) Calculate the correlation coefficient for the two variables.
2) Shuffle one of the variables, calculate this shuffled correlation coefficient, and record.

3) Repeat step two 1000 times.
4) When the observed correlation is positive: Find out how often the shuffled correlation coefficient is greater than the observed value. OR...
5) When the observed correlation is negative: Find out how often the shuffled correlation coefficient is less than the observed value.

The result is the *p*-value. Given the null model of no correlation, the *p*-value is the probability that a correlation coefficient as least as extreme as the observed value might occur by chance.

9.4.1.2 Example: Baseball Again

We return to the baseball payroll example, only this time using the correlation coefficient instead of the vector product sum. Under the null hypothesis of no linear correlation, the resampling (shuffled) distribution of the correlation coefficient centers, as expected, around 0 (Figure 9.8).

It rarely exceeds 0.5, so we can be safe in concluding that the value of 0.64 did not arise by chance and that the apparent correlation between wins and payroll is real.

9.4.1.3 Inference for the Correlation Coefficient: Formulas

Before computer-intensive resampling was available, a formula was needed to approximate the above-shuffled distribution. A modified sample statistic t (where n is the sample size and r is the sample correlation) follows Student's t-distribution with $n - 2$ d.f. (degrees of freedom) is given by:

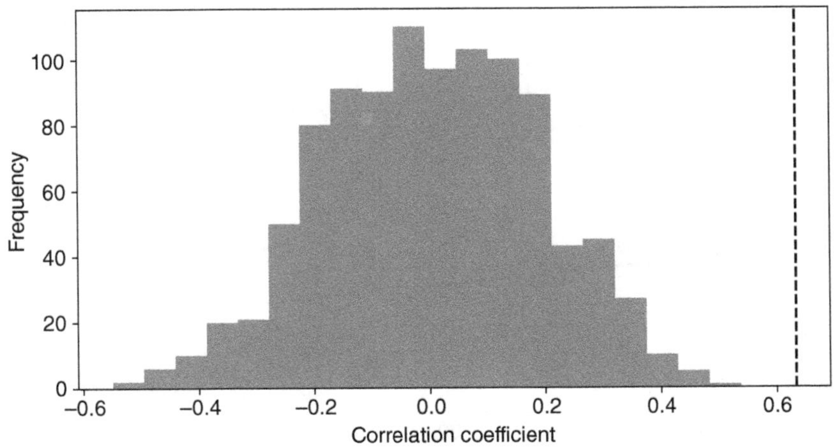

Figure 9.8 Resampling distribution of correlation coefficient for baseball under the null hypothesis.

$$t = \frac{r\sqrt{n-2}}{\sqrt{1-r^2}}$$

9.5 Correlation is not Causation

Correlation—even statistically significant correlation—does not imply anything about causation. Below, we present some examples of cases in which two variables are correlated, but causation is nonexistent.

9.5.1 A Lurking External Cause

In 1999, a University of Pennsylvania study found that infants who slept with a light on were more likely to develop myopia (near-sightedness) later in life. However, a later study found that the real cause was not the light but a genetic link to myopic parents. Lights in infants' rooms were more likely to be left on by myopic parents than non-myopic parents. The myopic parents needed the light to navigate.

In this case, the event "lights left on" was correlated with the development of myopia, but it was not the cause. When A causes both B and C, then B and C are correlated.

9.5.2 Coincidence

Consider murder rates in the 50 US states and the District of Columbia from 2021. The correlation between a state's murder rate and its position in the alphabetical

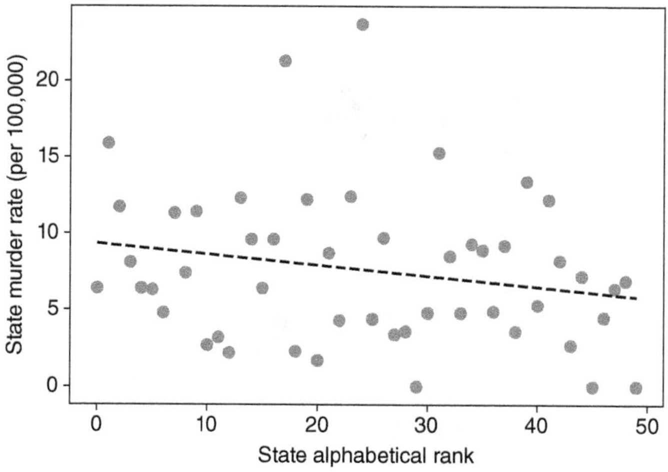

Figure 9.9 Murder rates and alphabetical order of states, $\rho = -0.28$.

order of states—Alabama being #1 and Wyoming being #50—is modestly negative at −0.21 (Figure 9.9). It is also statistically significant. Yet there seems no plausible reason why the two should be related, so we are inclined to think that this is due to coincidence.

Statistical significance is not a 100% guarantee that a relationship or finding is real rather than just coincidence. Significance testing reduces the chances of being fooled, but it does not eliminate them. The more you examine a given data set in hopes of finding something interesting, the greater your chance of finding something interesting that is actually due solely to chance.

Try It Yourself

Consider the following studies, and assess whether the reported correlation is probably part of a cause and effect relationship. Consider whether there is a reasonable theory to explain the correlation and whether some third factor might cause the correlation.

1) A web merchant finds a correlation between time spent on the website and money spent at checkout.
2) A medical study finds that elderly subjects who walk the fastest live the longest.
3) There is a positive correlation between income and education.
4) An exhaustive review of health records shows that higher consumption of zinc supplements is positively correlated with the scope of a person's social network.

9.6 Other Forms of Association

So far, we have been considering a relationship between x and y in which high values of x correspond with high values of y, or vice versa, on a consistent basis.

Now, consider these hypothetical data on the relationship between tax rates and tax revenue in the US (Figure 9.10).

At a zero tax rate, of course, no revenue is collected. As the tax rate, shown on the x-axis, increases, tax revenue, shown on the y-axis, also increases. However, at a certain point, the high tax rates make it increasingly profitable to employ tax avoidance schemes. Tax evasion increases. As a result, the revenue collected flattens out and begins to drop. At the point where the government seeks to take most of your money, the legal economy withers away. Most activity moves to the underground

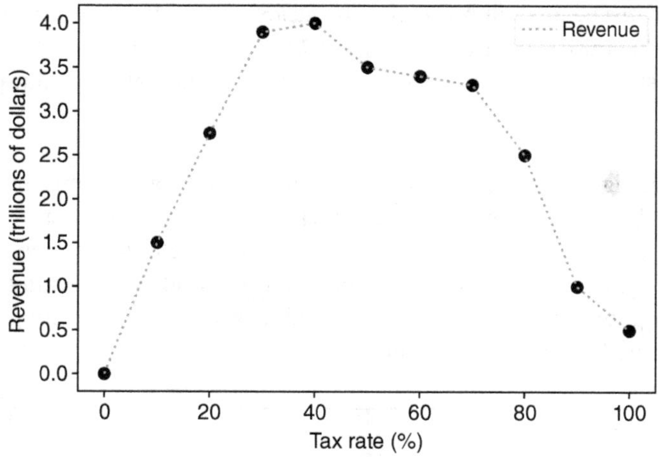

Figure 9.10 Hypothetical data on tax rates and revenue.

economy to escape taxes, and the tax revenues collected decline precipitously. The above data are hypothetical, but they are based in part on research concerning the "inflection point" where increasing rates do not yield additional revenue. There is much disagreement arising from differing political perspectives about the location of this point.

Clearly, there is a strong relationship between tax rates and revenue, but it will not show up clearly in the correlation coefficient for linear relationships because low revenue can be associated with both low tax rates and high tax rates. A nonlinear model is required to describe and investigate this relationship. We will briefly introduce nonlinear models in our prediction of binary outcomes (Chapter 12), but for now, we simply note that the absence of a strong positive or negative value for the correlation coefficient does not necessarily demonstrate the absence of an association.

9.7 Python: Correlation

9.7.1 Vector Operations

While Python has lists to represent vectors, it does not perform vector operations like addition, subtraction, or multiplication. However, we can implement them ourselves easily using list comprehensions. Here we illustrate multiplying a vector by a scalar (2), adding two vectors, and multiplying two vectors.

```
x = [1, 2, 3]
y = [2, 3, 4]
```

```
z = [2 * xi for xi in x]  # 2x = [2, 4, 6]
z = [xi + yi for xi, yi in zip(x, y)]  # x+y = [3, 5, 7] ①
z = [xi * yi for xi, yi in zip(x, y)]  # x*y = [2, 6, 12]
```

① The *zip* function returns the elements of the two lists in pairs as a `tuple`. This means it first returns `(1, 2)`, then `(2, 3)`, and finally `(3, 4)`. Each pair is assigned to the variables `xi` and `yi` in the list comprehension.

We can also implement the vector product sum from Section 9.3.

```
x = [1, 2, 3]
y = [2, 3, 4]
z = sum(xi * yi for xi, yi in zip(x, y)) ①
print(z)  # Output: 20
```

① As before, the *zip* function returns the pairs of elements from the two lists. We then multiply the elements of the pairs and sum them up.

We see that we can implement these operations in Python; however, it is not the most efficient way. The `numpy` package is a better choice. Here are the same operations using `numpy`.

```
import numpy as np
x = np.array([1, 2, 3])
y = np.array([2, 3, 4])
x = 2 * x # array([2, 4, 6])
z = x + y  # array([3, 5, 7])
z = x * y  # array([2, 6, 12])
```

The vector product sum can be written in either of the following ways.

```
z = np.sum(x * y)  # 20 ①
z = np.dot(x, y)  # 20 ②
```

① This statement first multiplies the elements of the two arrays in pairs. The elements of the resulting array are summed using the *np.sum* function.
② The *np.dot* function is a better way of calculating the vector product sum of the two arrays.

9.7.2 Resampling Test for Vector Product Sums

With the vector operations of the previous section, we can implement the resampling procedure from Section 9.3 and apply it to the *baseball_payroll.csv* dataset.

```
import pandas as pd
baseball = pd.read_csv("baseball_payroll.csv")

rng = np.random.default_rng(seed=321)

x = baseball["Average Payroll (Million)"]
```

```
y = baseball["Total Wins"]

resamples = np.array([np.dot(x, rng.permutation(y)) ①
                      for _ in range(100_000)])
observed = np.dot(x, y)

p_value = np.mean(resamples >= observed)
print(f"observed: {observed:.0f}")
print(f"resamples above observed: {np.sum(resamples >= observed)}")
print(f"p-value: {p_value}")
```

① In previous chapters, we used the *random.shuffle* command to randomize the elements of a list in place. The *rng.permutation* function is an alternative. It will return a randomized version of the vector y. The vector itself remains unchanged.

Output

```
observed: 668620
resamples above observed: 6
p-value: 6e-05
```

Only six out of 100,000 resamples have a larger vector product sum than the observed one. This results in a very low *p*-value. This is also clearly visible in the histogram of the resampled vector product sums in Figure 9.11a.

```
fig, ax = plt.subplots(figsize=(8, 4))
ax.hist(resamples / 1000, bins=30)
ax.axvline(observed / 1000, color="grey", linestyle="-")
ax.set_xlabel("Product sum (000)")
ax.set_ylabel("Frequency")
plt.show()
```

9.7.3 Calculating Correlation Coefficient

In this chapter, we learned that the correlation coefficient is a better measure of correlation than the vector product sum. The equation for the correlation coefficient *r* is:

$$r = \frac{\sum_{i=1}^{n} \left(X_i - \overline{X}\right)\left(Y_i - \overline{Y}\right)}{(n-1)\, s_x s_y}$$

Here is a function that implements this equation:

```
def correlation_coefficient(x, y):
    x = np.array(x) ①
    y = np.array(y)
    n = len(x)
    numerator = np.sum((x - np.mean(x)) * (y - np.mean(y)))
    denominator = (n - 1)*np.std(x, ddof=1)*np.std(y, ddof=1) ②
    return numerator / denominator
```

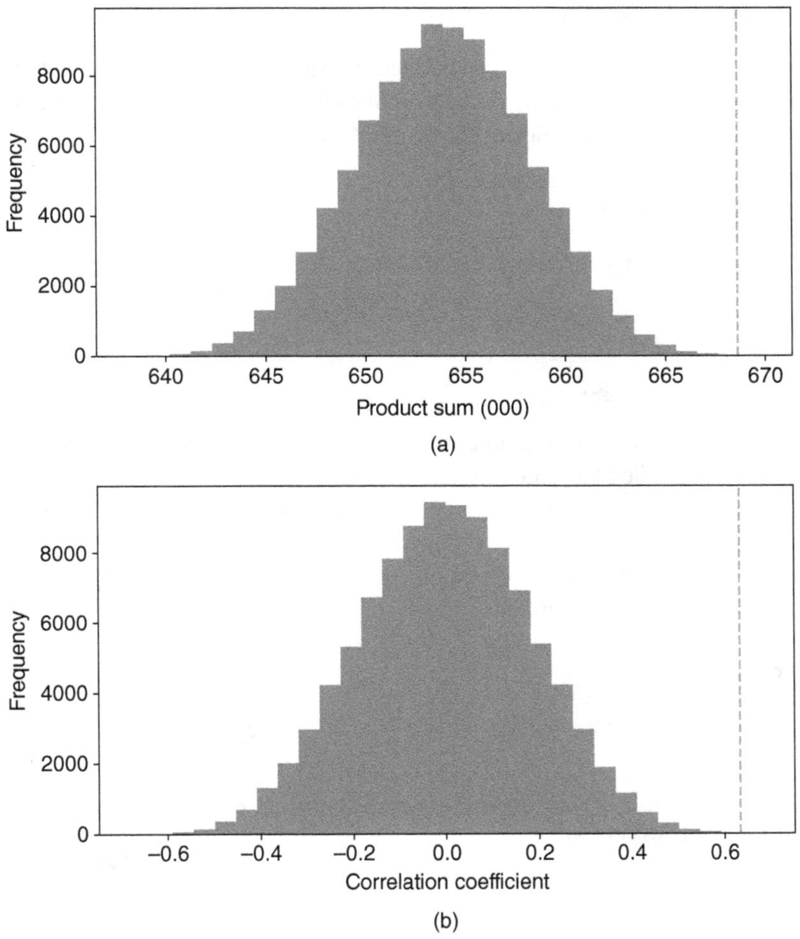

Figure 9.11 Distribution of resampled statistics for the Baseball dataset. In each graph, the observed statistic is shown as a dashed line. (a) Vector product sums and (b) correlation coefficient.

```
x = baseball["Average Payroll (Million)"]
y = baseball["Total Wins"]
r = correlation_coefficient(x, y)
print(f"correlation coefficient: {r:.3f}")
```

① This line and the next ensures that x and y are numpy arrays. Otherwise, this conversion would happen several times.

② We set the ddof parameter to 1 to calculate the sample standard deviation.

Output

```
correlation coefficient: 0.633
```

9.7.4 Calculate Correlation with numpy, pandas

Obviously, calculating the correlation coefficient is such a common operation that we find implementations in many packages. For example, the *np.corrcoef* function calculates the correlation coefficient for two vectors.

```
corrcoef_matrix = np.corrcoef(x, y)
print(corrcoef_matrix)
r = corrcoef_matrix[0, 1]
print(f"correlation coefficient: {r:.3f}")
```

Output

```
[[1.         0.63345733]
 [0.63345733 1.        ]]
correlation coefficient: 0.633
```

The function returns a matrix, not just a single value. This is because the *np.corrcoef* is more flexible and can calculate correlation coefficients between several vectors at once. The correlation coefficient between x and y is the value at position [0, 1] or [1, 0] in the matrix.

In pandas, we calculate the correlation coefficient between two columns of a DataFrame using the *corr* method.

```
corr_matrix = baseball[["Average Payroll (Million)", "Total Wins"]].corr()
print(corr_matrix)
r = corr_matrix.iloc[0, 1]
print(f"correlation coefficient: {r:.3f}")
```

Output

```
                           Average Payroll (Million)  Total Wins
Average Payroll (Million)                   1.000000    0.633457
Total Wins                                  0.633457    1.000000
correlation coefficient: 0.633
```

9.7.5 Hypothesis Tests for Correlation

In order to use resampling to test the hypothesis that an observed correlation coefficient is a result of random chance, we modify the procedure we developed for the vector product sum test.

```
rng = np.random.default_rng(seed=321)
resamples = np.array([np.corrcoef(x, rng.permutation(y))[0,1]
                      for _ in range(100_000)])
observed = np.corrcoef(x, y)[0,1]

p_value = np.mean(resamples >= observed)
print(f"observed: {observed:.3f}")
print(f"resamples above observed: {np.sum(resamples >= observed)}")
print(f"p-value: {p_value}")
```

Output
```
observed: 0.633
resamples above observed: 6
p-value: 6e-05
```

As before, only six out of 100,000 resamples have a larger correlation coefficient than the observed one. It is no surprise that we got the same *p*-value. Setting the random seed ensures that we create the same sequence of shuffled data and, therefore, a sequence of correlation coefficients that is perfectly correlated with the vector product sums and has the same distribution. We can also see in Figure 9.11b that the distribution is identical, with the exception of a linear transformation.

```
fig, ax = plt.subplots(figsize=(8, 4))
ax.hist(resamples, bins=30)
ax.axvline(observed, color="grey", linestyle="-")
ax.set_xlabel("Correleation coefficient")
ax.set_ylabel("Frequency")
plt.show()
```

9.7.6 Using the *t* Statistic

The modified sample statistic t can be calculated from the correlation coefficient r as follows.

$$t = \frac{r\sqrt{n-2}}{\sqrt{1-r^2}}$$

Using this equation, we calculate the t statistic for the observed correlation coefficient and estimate the *p*-value as follows.

```
from scipy import stats
r = observed
n = len(x)
t_value = r * np.sqrt(n - 2) / np.sqrt(1-r**2)
print(f"p-value {stats.t.sf(t_value, n-2):.2g}") ①
print(f"p-value {1 - stats.t.cdf(t_value, n-2):.2g}")
```

① The *stats.t.sf* survival function calculates the *p*-value for a given t statistic and degrees of freedom. It is the probability that the t statistic under the null hypothesis is larger than the observed one.

Output
```
p-value 8.6e-05
p-value 8.6e-05
```

The calculated *p*-value is very close to the one we determined using resampling.

The `scipy` function *stats.pearsonr* is an implementation of this approach. It returns the correlation coefficient and the *p*-value as a named tuple.

```
stats.pearsonr(x, y, alternative="greater")
```

Output

```
PearsonRResult(statistic=0.6334573297712217, pvalue=8.579436079820662e-05)
```

9.7.7 Visualizing Correlation

A great way of visualizing the correlation between two variables is a scatter plot. There are several examples of this in this chapter.

Another way of visualizing the correlation between several variables is a heatmap. We can use the *imshow* method of an axis object to create a heatmap.

```
import matplotlib.pyplot as plt

housing = pd.read_csv("boston-housing-large.csv")
fig, ax = plt.subplots()
ax.imshow(housing.corr(), cmap="coolwarm", vmin=-1, vmax=1) ①
ax.set_yticks(range(housing.shape[1])) ②
ax.set_yticklabels(housing.columns)
plt.show()
```

① The *imshow* method creates a heatmap from the correlation matrix. The colors of the heatmap are defined using a colormap (`cmap`). We set `vmin=-1` and `vmax=1` to make sure that the colors are centered around zero and that the full range of possible correlation coefficients is mapped onto the colors.

② The *set_yticks* method sets the ticks on the *y*-axis. We do this to ensure that each column name is shown in the final figure. The labels are set to the column names using the *set_yticklabels* method.

The result is shown in Figure 9.12a. The heatmap shows the strength of the correlation between the variables. The darker the color, the stronger the correlation. The color indicates the sign of the correlation.[3]

Figure 9.12b shows a variation. The colored rectangle patches are replaced by ellipses that represent the sign of the correlation by color and direction and the strength of the correlation by the size of the ellipses. The code for Figure 9.12b is too complex to include here. You can find it in the Jupyter Notebook for this chapter.

---

3 In the print version of this book, the heatmap is shown in grey. Check the accompanying Jupyter Notebooks for colored versions of the heatmaps and for the code to create the variation using ellipses.

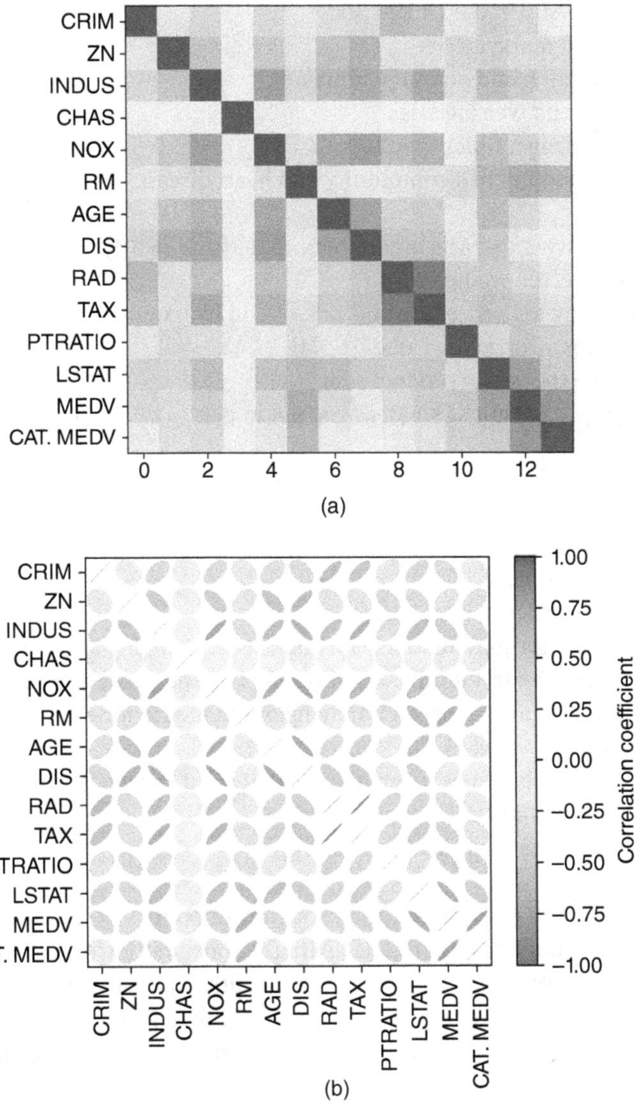

Figure 9.12 Visualizing correlation between two variables. (a) Heatmap and (b) variation of heatmap with ellipses showing the strength of the correlation.

Exercises

9.1 In the following situations, what form of association would you expect to see between the variables—linear, nonlinear, or none at all? If linear, would it be positive or negative? And would you expect there to be a causal relationship?

a) Income and obesity in the United States
b) Advertising and market share
c) Cement production and potato production
d) Website traffic and website sales
e) Air traffic (passenger miles) and month
f) Alcohol consumption in a community and heart disease

9.2 Consider the following two sets of numbers: [1 2 3] and [6 5 4]
a) Calculate the vector product sum
b) Repeat the following procedure and calculation by hand five times:
   i) Randomly rearrange the values [6 5 4]
   ii) Recalculate the vector product sum
c) Did you ever get a result as small as you got in part (a)?

9.3 Activity in the portions of the brain connected with social perceptions was tracked for a sample of university students, and the number of Facebook friends was recorded for the same sample. Review the data in *brain-facebook.csv*, then:
a) Plot the two variables in a scatterplot.
b) The metric used for gray matter (GM) density is complex and it is not necessary to understand it fully. However, look at the *y*-axis and make a guess about the final step used in arriving at the metric for each person (*Hint*: review Section 5.4.1).
c) Calculate the correlation between brain activity and Facebook friends.
d) Test whether it is statistically significant.
e) Discuss: What is the direction of causality?

9.4 In the 1950's, in the United States, the incidence of polio was on the rise. Disease rates were found to be correlated with the consumption of ice cream (more ice cream, more polio), and some medical authorities advised parents not to feed ice cream to their children. Can you think of an external factor that might be correlated with both ice cream consumption and polio incidence?

9.5 For the following exercise, we use the *pulse.csv* dataset.
a) Load the data and determine the correlation coefficients between all numerical variables.
b) Visualize the correlation matrix as a heatmap.
c) Show scatterplots for the pairs of variables with correlation coefficients.

10

Regression

With correlation, all we can measure is the relative strength of a linear association and whether it is statistically significant. In this chapter, we move on to regression and quantifying *how much* an outcome variable changes as you change a predictor variable or variables. After completing this chapter, you should be able to:

- Estimate, by eye, a trend line from a two-dimensional scatterplot
- Explain the derivation of a two-dimensional least-squares regression line
- Interpret the meaning of a regression line
- Use a regression line to predict values
- Interpret a residual plot

With regression, we can model that association in linear form and predict values of Y given values of X. The simple form of a linear regression model is as follows:

$$y = ax + b$$

We read this as "y equals a times x, plus a constant b." You will note that this is the equation for a line with slope a and intercept b. The value a is also termed the coefficient ("coef" in software output) for x (Figure 10.1). Regression is one of a number of statistical and machine learning algorithms used to predict some target variable of interest (y) on the basis of predictor variables (one or more x variables).

Definition: Algorithm An algorithm is a procedure or set of rules for solving a problem, typically specified with sufficient precision that they can be followed unambiguously by a computer. The resampling procedures we have been following in this book are good examples of algorithms.

Statistics for Data Science and Analytics, First Edition. Peter C. Bruce, Peter Gedeck, and Janet Dobbins.

Companion website: www.wiley.com/go/Wiley_Statistics_for_Data

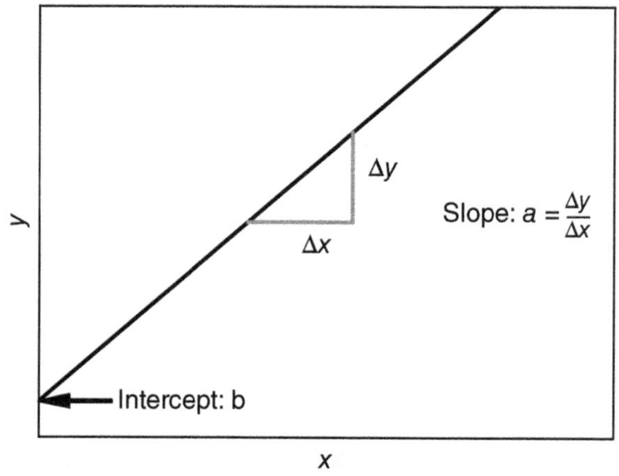

Figure 10.1 Slope and intercept of a line.

10.1 Finding the Regression Line by Eye

Using the baseball payroll example we introduced in the last chapter and assuming correlation exists between the payroll amount in million and the number of wins over three seasons, can we predict wins based on a given payroll amount?

Based on Figure 10.2, it appears that an increase in payroll generally predicts an increase in wins. Suppose we ask the question, "How many wins can be expected over a three-year period with a payroll of $130 million dollars?" It might be possible to arrive at a reasonable answer based on the scatterplot.

A better first step might be to find a line that best represents the data. This is sometimes called a "line of best fit" or a "trend line." An eyeball estimate of such a line appears in Figure 10.3.

Assuming that the trend line is reasonable, one can estimate approximately 258 wins over three seasons with a payroll of $130 million. The word "approximately" is important. Other factors—including, of course, chance—affect performance, so relatively few teams fall *exactly* on the trend line. The difference between an actual *y* value and the predicted *y* value is called a residual or error.

Definition: Residual Residuals are the differences between the actual *y* values and the values predicted by the trend line or linear regression equation. In a world of perfect prediction, all residuals would be zero, indicating that no error exists between the prediction and the true value. In reality, this never happens, and there is always some error. The key to finding the optimum regression line is to minimize the error in the residuals.

$$\text{Residual} = Y - \hat{Y}$$

where Y is the actual value and $\hat{Y}$, pronounced "Y-hat," is the predicted value.

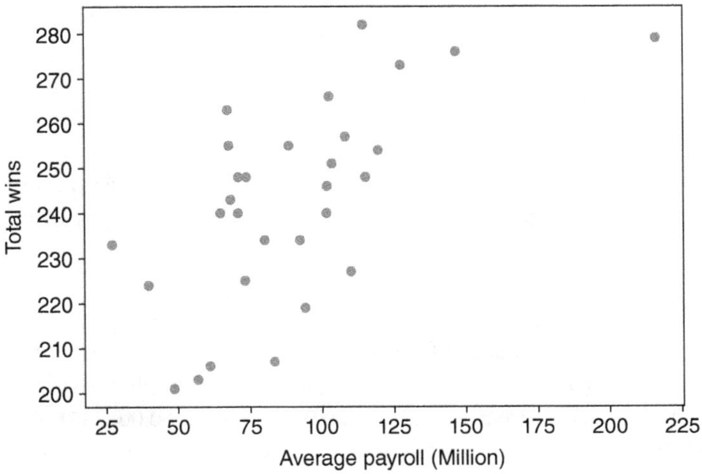

Figure 10.2 Payroll vs. total wins 2006–2008.

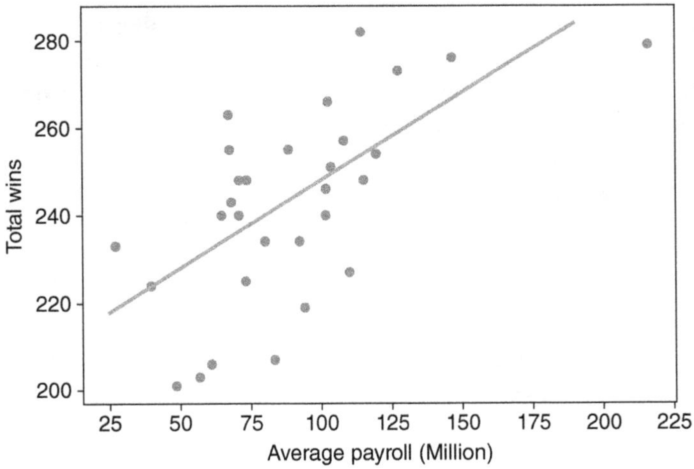

Figure 10.3 Estimated trend line, drawn by eye.

Definition: Error In statistics, the term "error" does not mean "mistake." Rather, it simply means the difference between predicted and actual values. It is the same thing as the residual. Error in a positive direction is just as bad as error in a negative direction, and to keep positive and negative errors from canceling each other out, we use either absolute error or, more commonly, squared error.

Now, we can add an error term, *e*, to the regression model. The equation for the regression line is now:

$$y = ax + b + e$$

We read this as "*y* equals *a* times *x*, plus a constant *b*, plus an error *e*."

The error can result from the effect of an additional variable that we did not include, or it might simply be a random error resulting from unaccountable causes.

10.1.1 Making Predictions Based on the Regression Line

Let's go one step further. If an equation for the regression line can be found, the estimation of wins based on payroll becomes easier. Based on the graph, the regression line goes through or very close to the points (30.00, 220) and (60.00, 232). The slope of this line is 0.40, which means that, in general, as payroll increases by (say) $10 million, the number of wins increases by 4. For this example, the equation for the estimated regression line is

$$y = 0.40x + 208$$

The variable *x* is the predictor. In this case, *x* is the payroll in millions of USD. The *y* variable represents the predicted number of wins. The constant 208—the *y*-intercept—is the value the equation takes when *x* represents zero payroll (Figure 10.3). A prediction at $0 is meaningless, of course, since it is not possible in Major League Baseball to field a team without paying the players.

Using $130 million as the predictor, $0.40 \times 130 + 208 = 260$ wins, which is very close to our eyeball estimate. Predicting values within the limits of the data is called interpolation. Predicting values outside the limits of the data is called extrapolation.

What conclusions could you draw from extrapolation in this case?

Placing a trend line by eye based on an aesthetic best fit may work in some cases, but a more scientific method is needed to cover all situations. In particular, when we move to multiple predictors, a two-dimensional plot doesn't help.

10.2 Finding the Regression Line by Minimizing Residuals

We will use Python to calculate the trend line by finding the line that minimizes the error.

Figure 10.4 Minimizing residuals.

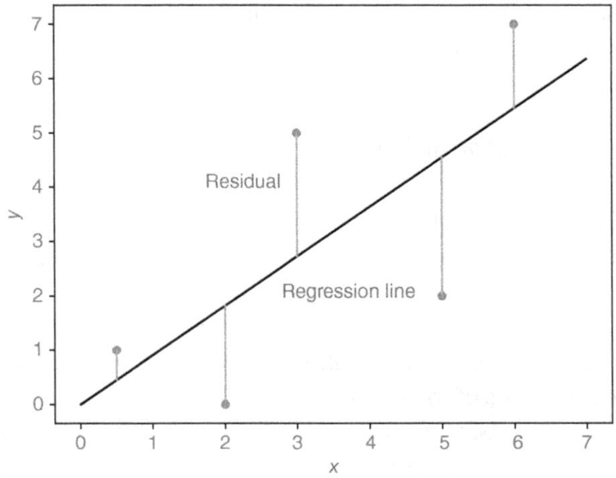

In other words, the computer will find the equation of the line that minimizes the sum of the error terms—the residuals. Figure 10.4 shows the residuals as the vertical lines between the regression line and the observations.

In practice, the mathematics of linear regression does not minimize the absolute residual error. Instead, for mathematical convenience, it minimizes the squared residual error. This procedure is called Least Squares Regression.

Definition: Least Squares Regression The least squares regression line for *x*–*y* data is the line that minimizes the sum of the squared residuals between the actual *y*-values and the *y* values that are predicted by that line.

The least squares regression line for the baseball payroll data shown above is:

$$y = 0.39047x + 207.4793$$

10.2.1 The "Loss Function"

The least squares is a specific example, of a *loss function* for a statistical or machine learning predictive model. (In this chapter, we are looking at a linear model for numerical data, but there are a number of different models that do not assume a linear relationship.) A predictive model like regression requires some metric that measures how well it does its prediction. For example, if you are predicting binary outcomes, one possible loss function is accuracy—the percent of cases predicted correctly. The loss function we are using here for regression is the sum of squared errors. In this case, the loss function of the sum of squared errors is used to find the regression line. Loss functions can also be used to compare the performance of different statistical or machine learning models applied to the same data. The square root of mean squared error (RMSE) is commonly used for this purpose with numerical data.

10.3 Linear Relationships

Linear relationships are powerful. If a linear relationship exists between two variables, then it becomes possible to predict the unknown value of one variable based on the known value of another.

Some linear relationships are obvious. Figure 10.5 illustrates the Delta Wire example from the correlation chapter, with a regression line added. As the number of hours of training increases, so does the productivity. The linear relationship appears to be very strong, and the line almost draws itself.

The least squares regression line equation is (Figure 10.5):

$$y = 5.093445x + 70,880.25$$

When an obvious linear relationship does not appear to exist, what can be done?

10.3.1 Example: Workplace Exposure and PEFR

Consider the data shown in Figure 10.6. How and where would a trend line be placed? No linear relationship is obvious. The least squares regression line calculated by Python can be viewed in Figure 10.7.

The least squares regression line is:

$$y = -4.18458x + 424.5828$$

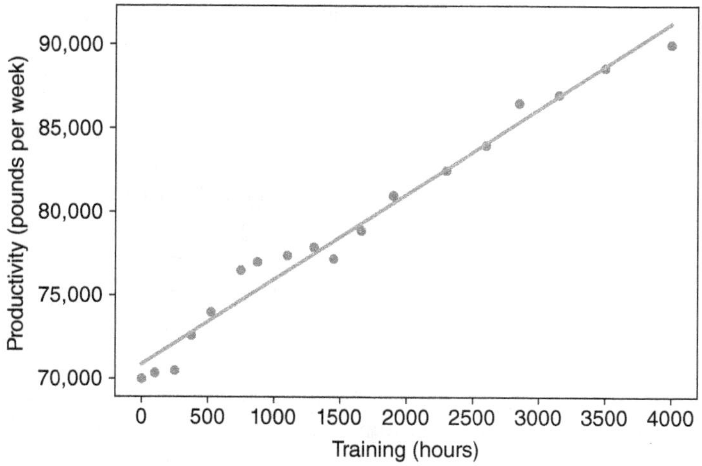

Figure 10.5 Delta Wire hours of training and productivity.

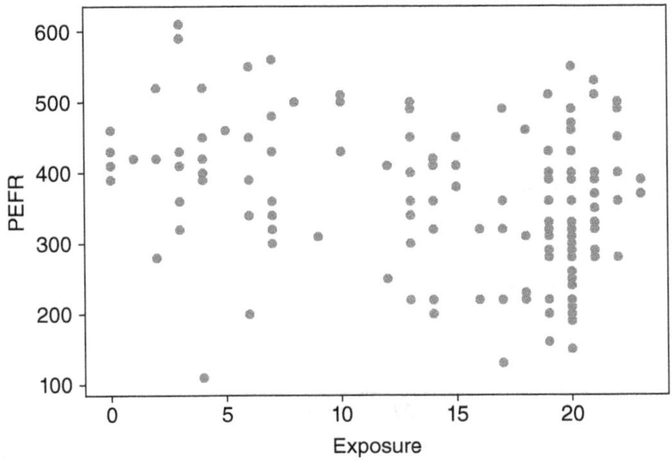

Figure 10.6 Pulmonary capacity (PEFR) and exposure to cotton dust (years).

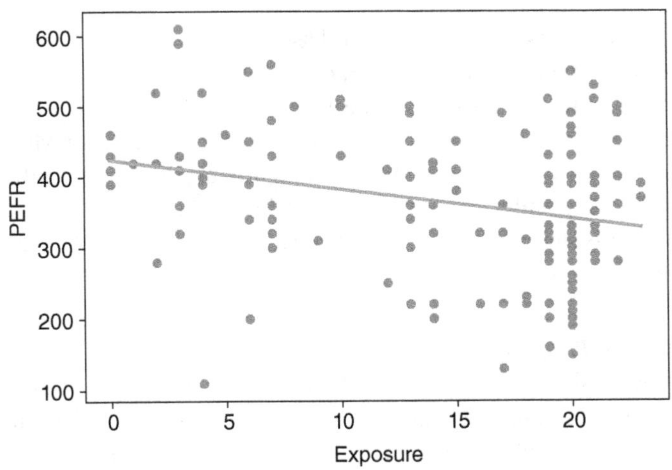

Figure 10.7 Trend line with negative slope.

It appears that there may be a slight negative relationship between exposure to cotton dust and pulmonary effusion rates. This means that as the years of exposure increase, the pulmonary capacity decreases, albeit slowly. But the data are all over the place—how much confidence can we have in this linear relationship?

10.3.2 Residual Plots

One method for assessing the predictive ability of a trend line is to look at a plot of residual values.

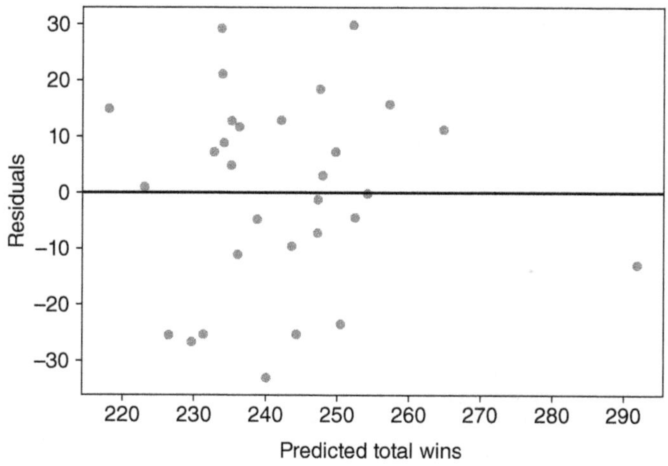

Figure 10.8 Payroll residual plot.

Figure 10.8 illustrates the residual plot for the baseball payroll data. At first glance, it appears from the plot that there is a large amount of error. To interpret the amount of error and, indirectly the confidence we have in our model equation, it is useful to compare the scale of the residuals to the scale of the number of wins in the original data. The residual values range from approximately +30 to −30, a range which is somewhat smaller than the range of wins (202 to 283). The fact that, in general, the error is smaller than the variability in the data reflects the fact that the regression equation is useful.

10.3.2.1 How to Read the Payroll Residual Plot

The x-axis is the fitted value, i.e. the predicted number of wins for a team. (Sometimes, for simple regression with a single predictor, the plot is constructed with the x-axis showing the independent or predictor variable.) The y-axis is the residual between the predicted number of wins and the actual number of wins. Consider the left-most point. The x-coordinate of this point is the predicted number of wins for what was then called the Florida Marlins and is now called the Miami Marlins. Our regression equation predicted about 218 wins, and yet the Marlins won 233 games. Thus, the residual—actual wins – predicted wins—is 15, which is the y-coordinate of the point.

Figure 10.9, the Delta Wire Training Residual Plot, has a residual error range of approximately −1600 to +1800. This seems high, but consider that the productivity range is from 70,000 to 90,000! The residual error is relatively small, and we can see this in the linear nature of the data shown in Figure 10.5.

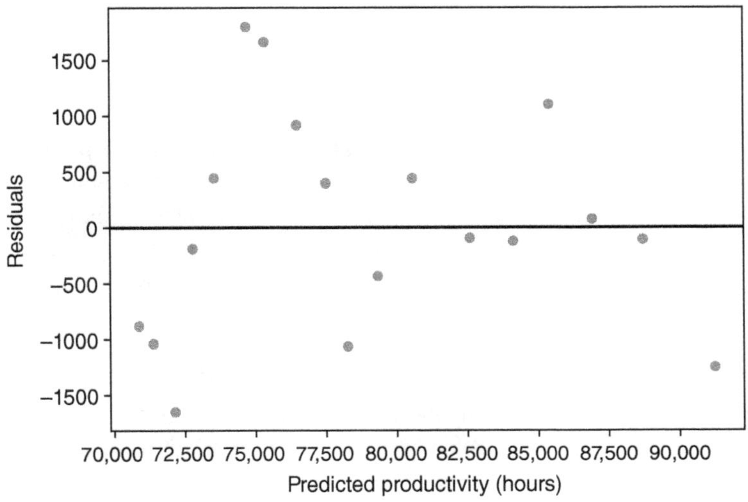

Figure 10.9 Delta Wire training residual plot.

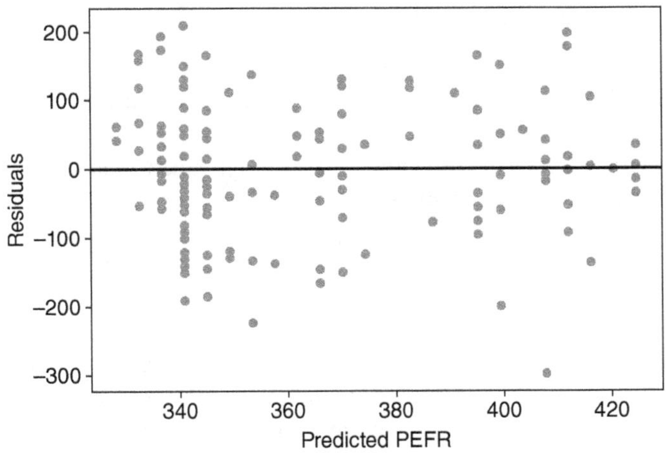

Figure 10.10 PEFR residual plot.

At first glance, the residuals in the PEFR plot in Figure 10.10 appear to be similar to the previous plots. However, the residual error ranges from −300 to +210, and the actual data ranges from 110 to 610. The residual range is as large as the data range! This indicates that either the regression equation fit isn't very good or there isn't much of a relationship between the two variables. In this case, from looking at the original data scatterplot in Figure 10.6, it doesn't appear that a strong relationship exists.

While the relationship is not strong, it still could be

1) Statistically significant, and
2) Useful

The utility can be measured in terms of human health. Although respiratory function varies widely among individuals, an improvement in overall average respiratory function is still very valuable. Most improvements in health conditions and treatments are not sweeping and revolutionary. If they help only 10% of the population, they are still very meaningful, but they will not necessarily show a dramatic statistical picture.

Note: There is probably some inherent difference among individuals with respect to PEFR, which adds noise to the picture. A plot of change in PEFR (which was not available), rather than PEFR itself, would control for the great differences among individuals and probably show a stronger relationship.

10.4 Prediction vs. Explanation

Regression is used in two different contexts:

1) In research studies, to quantify and confirm relationships between predictor (independent) variables and outcome (dependent) variables. It is the *relationship* that is of interest.
2) In predictive modeling, to predict individual values for outcome variables. It is the *individual predictions* that are of interest.

10.4.1 Research Studies: Regression for Explanation

In the examples presented in Chapter 9, we spoke about using linear regression to predict values, but we were interested in more than individual values: we wanted to learn something about how strong and reliable the relationship was and how sound the regression model was as an explanation for the phenomenon being studied.

In the baseball payroll example, the user of this model might be a management team in baseball seeking to allocate budget for payroll. A simple prediction of "X wins if you spend Y dollars" might be of interest, but they would want to know all about the relationship and how well it explains the data.

- Is it a strong relationship? Does a boost in spending produce a big boost in performance?
- Is it linear? Or are there ranges where spending more might have no effect?
- Is it reliable? Are the data clustered tightly along the regression line, giving you confidence in the relationship, or are they all over the place?

In the example relating cotton dust exposure to pulmonary function, there is probably almost no interest in an individual prediction. The purpose of the research is to inform public policy and industry health and safety standards, which requires an evaluation of the data as a whole.

Regression has long been used in this way—to characterize and explain an overall relationship, typically as part of a research study. Data may be sufficiently scarce that you want to use all of it in the study. Consumers of the study (other researchers, public policymakers, regulators, etc.) will want to know how well the regression model fits the data, whether relationships are statistically significant, what the confidence intervals are, and how sound your overall analysis is.

10.4.2 Assessing the Performance of Regression for Explanation

The traditional machinery for assessing regression assumes all of the available data are used in the study and in fitting a regression model. So the model is typically assessed on the basis of how well it does with those data (Figure 10.11). In statistical software, including Statsmodels, the default reporting mode is to calculate the model evaluation metrics using the data that was used to fit the model.[1]

A variety of metrics are used to assess the performance of regression for explanation, including:

1) *P*-values to measure the statistical significance of the coefficients and constant
2) The *F* statistic to measure the significance of the overall model (see ANOVA chapter)
3) Standard errors (SE) for the coefficients and constant

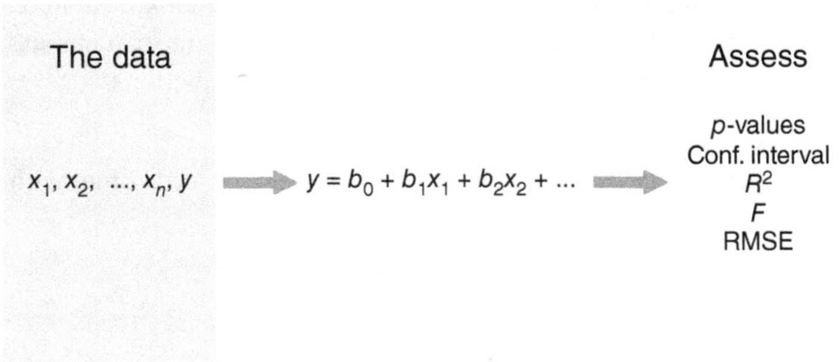

Figure 10.11 Assessing the performance of regression.

1 Evaluating regression models using holdout data, as is done for prediction models, is an increasingly common practice for research projects where there is enough data to support this. This may involve ignoring the default software reporting metrics. See the section on regression for prediction.

4) Confidence intervals for the coefficients and constant
5) Overall R^2 to measure the proportion of the variation in the data that is explained by the model
6) RMSE, or root-mean-squared-error (see Section 10.4.4)

Some metrics are used to assess the overall validity of the model (F statistic) and its strength (R^2). Others assess the reliability of the individual predictor variable coefficients and the constant in light of possible chance variation. *P*-values answer the question "is this predictor coefficient or constant significantly different from what we might get by chance?" Confidence intervals provide a range for the likely coefficient value. Standard errors (see Section 7.7) embody in a single number the potential chance variation in the coefficient or constant.

10.4.3 Big Data: Regression for Prediction

The advent of "big data" added a new paradigm for using regression and similar models and gave rise to the field of "data science." "Big data" are data typically produced by ongoing processes such as e-commerce and social media interactions, with many variables and vast numbers of records—and that's just structured data that are stored in databases. "Unstructured" data, often in the form of text and images (Twitter feeds, Facebook postings, tech support records, email, e-commerce documents, weblogs, etc.), constitute an even larger source of continually generated data. Amazon alone requires 1.4 million servers to store its data, and the pace of data generation is accelerating. It is estimated that 90% of all data was generated in the last two years. The techniques needed to analyze text and image data are complex and lie beyond the scope of this text, but they rest on the foundation of statistical and machine learning models that we are looking at here.

Organizations with access to big data (which is most organizations) can use statistical modeling techniques like regression to make valuable predictions.

- How much will a customer spend?
- How much will an insurance company have to pay in claims for a customer with certain characteristics?
- Is an email spam?
- Which online ad is a customer most likely to click on?
- Will a borrower default?

In all these cases, the main goal is to make individual predictions. Understanding and explaining the overall relationship is usually a secondary issue, though it can be useful as part of the "due diligence" around a statistical or machine learning model.

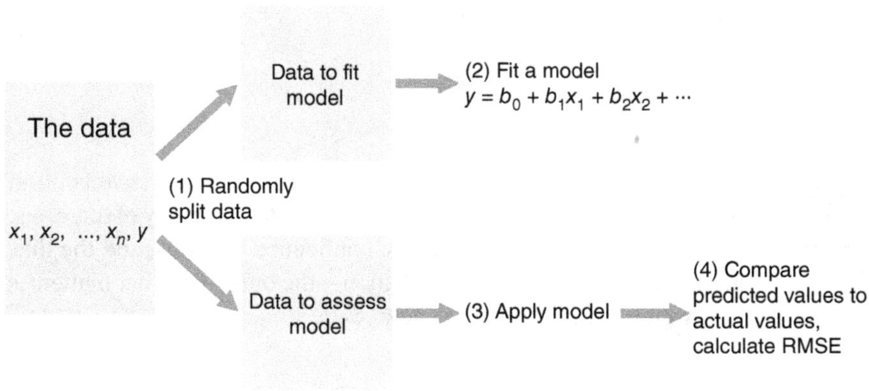

Figure 10.12 Assessing the performance of regression for prediction.

10.4.4 Assessing the Performance of Regression for Prediction

Since the data are plentiful, they can be split up into separate partitions, and the accuracy of predictions can be tested directly with "holdout" data.

Definition: Training and Holdout Samples Records used in building regression models can be randomly partitioned into training data and holdout data (Figure 10.12).

Training data are the records used to fit the model. Usually, they constitute 50%–70% of the data.

Holdout data are the records to which the model is applied to see how well it does.

There are many ways of dividing (partitioning) datasets using Python. We will cover this in more detail in Section 10.5.4. Once we have the partitions, one of the partitions (the "training" partition) can be used to fit a model. The model is then used to "score" the other partition (the "holdout" partition), meaning that it uses the model to calculate the predicted y value for each record (this process is also called "inference"). The residual, or error, between the predicted y value and the actual y value in the holdout data, then goes into calculations that measure the overall accuracy of the model.

One overall measure of error is "Root Mean Squared Error" or RMSE, using the residuals (predicted vs. actual values). It is calculated by squaring the residuals, taking the average, and then finding the square root. You can think of this as the "typical" error in each predicted value.

Definition: Root Mean Squared Error

$$\sqrt{\frac{1}{n}\sum_{i=1}^{n}(\hat{y}_i - y_i)^2}$$

Most statistical software will automatically produce regression output with a variety of metrics and diagnostics. Note that many of these, e.g. *p*-values, goodness-of-fit statistics, confidence intervals, and the like, are usually not relevant for prediction—the only thing that matters is how well the model predicts "new" data.

10.5 Python: Linear Regression

10.5.1 Linear Regression Using Statsmodels

We already used the `statsmodels` package to perform an ANOVA in Chapter 8. This is just one of the many statistical models and tests the `statsmodels` package implements. `Statsmodels` can do more; it can also perform time series analysis, generalized linear models, and many other statistical models. In this chapter, we will use it to build linear regression models.

The `statsmodels` package is loaded, usually, as follows:

```
import statsmodels.api as sm
import statsmodels.formula.api as smf
```

The `statsmodels.api` module contains the core functionality of the package. This includes statistical models like linear regression. However, it is often more convenient to use the formula interface to these models that is available in the `statsmodels.formula.api` module. This interface allows us to specify the model using a formula string, similar to the formula strings used in R and other statistical software. Let's look at an example using the *delta-wire.csv* dataset. We build a linear regression model to predict productivity from the training time. The formula interface is used to specify the model. The formula string `"productivity ~ training"` specifies that we want to predict the productivity from the training time. The *smf.ols* function is used to fit the model. The function name *ols* stands for ordinary least squares, which is the method used to fit the model. This is how it is used:

```
import pandas as pd
delta_wire = pd.read_csv("delta-wire.csv")
formula = "productivity ~ training" ①
model_definition = smf.ols(formula, data=delta_wire) ②
model = model_definition.fit() ③
```

① The formula string specifies the model. To the left side of the tilde is the dependent variable *y*, and to the right side is the independent variable *x*.
② The *smf.ols* function is called with the formula and the dataset. It returns the model instance `model_definition`. At this point, the model is not yet fitted to the data.
③ We call the *fit* method of the model instance to fit the model to the data. The result is the fitted model assigned to the variable `model`.

We can get a summary of the model using the *summary* method of the fitted model. The output of the *summary* method is shown in Figure 10.13.

```
print(model.summary())
```

There is a lot of information in Figure 10.13. The most important part is the table with the coefficients. We can see that the intercept is about 70,880. This is the productivity of an employee without any training. The coefficient of the training time is about 5.09. This means that for every hour of training, the productivity increases by 5.09. The r^2 value of the model is 0.976.

```
                            OLS Regression Results
==============================================================================
Dep. Variable:           productivity   R-squared:                       0.976
Model:                            OLS   Adj. R-squared:                  0.975
Method:                 Least Squares   F-statistic:                     662.3
Date:                Sat, 17 Feb 2024   Prob (F-statistic):           1.90e-14
Time:                        19:08:01   Log-Likelihood:                -148.92
No. Observations:                  18   AIC:                             301.8
Df Residuals:                      16   BIC:                             303.6
Df Model:                           1
Covariance Type:            nonrobust
==============================================================================
                 coef    std err          t      P>|t|      [0.025      0.975]
------------------------------------------------------------------------------
Intercept   7.088e+04    394.546    179.650      0.000       7e+04    7.17e+04
training       5.0934      0.198     25.735      0.000       4.674       5.513
==============================================================================
Omnibus:                        0.348   Durbin-Watson:                   0.812
Prob(Omnibus):                  0.840   Jarque-Bera (JB):                0.495
Skew:                           0.207   Prob(JB):                        0.781
Kurtosis:                       2.301   Cond. No.                     3.32e+03
==============================================================================
```

Figure 10.13 Output of the *summary* function of the fitted model.

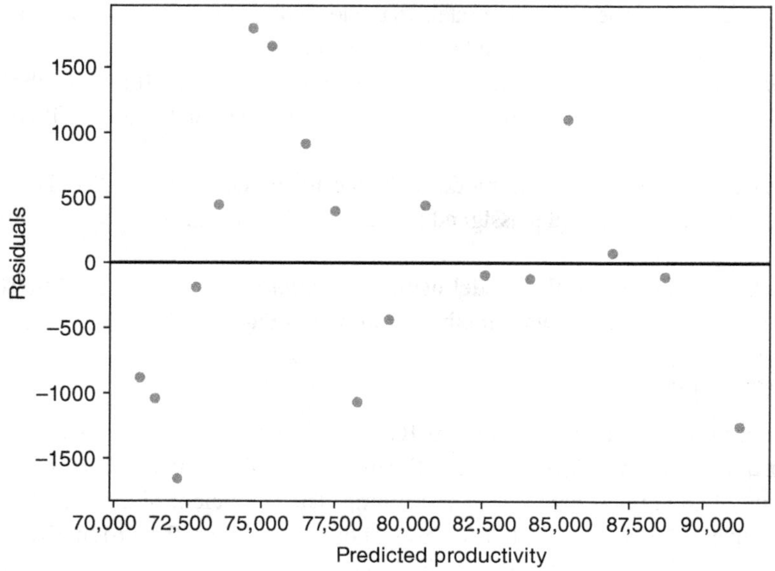

Figure 10.14 Residual plot of the fitted model.

The fitted model also contains information about the residuals (`model.resid`) and the predicted outcome (`model.fittedvalues`), which allows us to create the residual plot shown in Figure 10.14.

```
df = pd.DataFrame({"predicted": model.fittedvalues,
"residuals": model.resid})
ax = df.plot.scatter(x="predicted", y="residuals")

ax.set_xlabel("Predicted productivity")
ax.set_ylabel("Residuals")
ax.axhline(0, color="black")
plt.show()
```

The fitted model can be used to make predictions using the *predict* method. The code below shows how to use the model to predict the productivity for training times of 1230 and 2390 hours.

```
prediction = model.predict({"training": [1230, 2390]}) ①
print(prediction)
```

① Because we used the formula, we need to call the *predict* method with a data structure that allows us to access the independent variables by name. Here, we use a dictionary with the variable name as a key and the values as a list. An alternative would be to use a `pandas DataFrame`. The result of the prediction is a `pandas Series` with the predicted values.

Output

```
0    77145.189207
1    83053.584830
dtype: float64
```

The formula interface automatically added an intercept to the model. However, what if we wanted to create a model with an intercept of 0? In this case, we can modify the formula as follows:

```
formula = "productivity ~ training - 1"
formula = "productivity ~ training + 0"
```

Both are equivalent and will create a model without an intercept. There is a lot more to know about formulas in `statsmodels`. For example, you can specify interactions between variables, use the *Q* function for column names with spaces, or use the *C* function to specify categorical variables. We will see more of this in this chapter. You can also find a wealth of information in the `statsmodels` documentation.

10.5.2 Using the Non-formula Interface to `statsmodels`

The formula interface is convenient for simple models. However, it is not always necessary to specify the model using a formula. As we will see in the next chapter, we can extend linear regression to more than one independent variable, and if their number gets very large, there will be no advantage in using formulas. In these cases, we can use the non-formula interface for `statsmodels`. In the non-formula interface, you provide the dependent and independent variables as arguments to the model. The dependent variable is called the endogenous variable, and the independent variables are called the exogenous variables. These terms come from econometrics, where the endogenous variables are the variables that are determined within the model, and the exogenous variables are the variables that are determined outside the model. Let's build the same model as before using the non-formula interface. A first attempt might look like this:

```
import statsmodels.api as sm
import pandas as pd
delta_wire = pd.read_csv("delta-wire.csv")
X = delta_wire["training"]
y = delta_wire["productivity"]
model = sm.OLS(y, X)
results = model.fit()
print(results.summary())
```

Here is the partial output of the `print(results.summary())` statement:

```
                         OLS Regression Results
==============================================================================
.....
==============================================================================
                 coef    std err          t      P>|t|      [0.025      0.975]
------------------------------------------------------------------------------
training      33.5179      5.182      6.468      0.000      22.584      44.451
==============================================================================
.....
==============================================================================
```

The output shows that the model has only one coefficient, which is the coefficient of the training time; there is no `Intercept`. This is not what we want. The problem is that the *sm.OLS* function expects the design matrix to be the second argument. The design matrix is a matrix that contains the independent variables, and the constant intercept can be one of them. The `statsmodels` package provides the *add_constant* function to do just this (it does this by adding a column of ones to the design matrix). This column is used to fit the intercept of the model. Let's see what happens if we increase the design matrix by adding the constant term.

```
X = delta_wire["training"]
y = delta_wire["productivity"]
X = sm.add_constant(X)
model = sm.OLS(y, X)
results = model.fit()
print(results.summary())
```

Output

```
                         OLS Regression Results
==============================================================================
.....
==============================================================================
                 coef    std err          t      P>|t|      [0.025      0.975]
------------------------------------------------------------------------------
const       7.088e+04    394.546    179.650      0.000       7e+04    7.17e+04
training       5.0934      0.198     25.735      0.000       4.674       5.513
==============================================================================
.....
==============================================================================
```

The resulting model is identical to the model we built using the formula interface.

10.5.3 Linear Regression Using `scikit-learn`

The `scikit-learn` package is a powerful package for machine learning. In contrast to `statsmodels`, `scikit-learn` is not a package for statistical analysis. It is a package for machine learning. This means that `scikit-learn` is more focused on predictive modeling than on statistical inference. It implements a wide variety of machine-learning algorithms for regression, classification, clustering, and dimensionality reduction. It also provides tools for preprocessing data, model selection, and evaluation. We will only scratch the surface of what `scikit-learn` can do in this chapter and show how to use it to build a linear regression model.

Here is how to build the linear regression model using `scikit-learn`:

```
import pandas as pd
from sklearn.linear_model import LinearRegression ①
delta_wire = pd.read_csv("delta-wire.csv")
X = delta_wire[["training"]]
y = delta_wire["productivity"]
model = LinearRegression() ②
model.fit(X, y) ③
print(f"Model intercept: {model.intercept_}") ④
print(f"Model coefficients: {model.coef_}")
```

① The `scikit-learn` package is loaded as the module `sklearn`.
② The `LinearRegression` class is used to define a linear regression model. At this point, the model is not yet fitted to the data.
③ The *fit* method fits the model to the data.
④ We can access the intercept and coefficients using the `intercept_` and the `coef_` attributes. Both names end in an underscore. This is a convention in `scikit-learn` to identify attributes that are set during the fitting process.

Output

```
Model intercept: 70880.25246964331
Model coefficients: [5.0934445]
```

All predictive models in `scikit-learn` have the same API (application programming interface). This means they all have a *fit* method that takes the dependent and independent variables and fits the model to the data. Another method that is shared between all predictive models is the *predict* method. This method is used to make predictions using the fitted model. The code below shows how to use the fitted model to make predictions for the training times 1230 and 2390.

```
new_data = pd.DataFrame({'training': [1230, 2390]})
prediction = model.predict(new_data) ①
print(prediction)
```

① The *predict* method takes a matrix with the independent variables as an argument. Here, we use a dataframe that contains the same column name as the training data. It is possible to call this function using a list of lists like `[[1230], [2390]]`. If you do this, `scikit-learn` will print a warning that the model was trained with feature names. It is better, in this case, to use a dataframe like we've done here. The result of the prediction is a `numpy` array with the predicted values.

Output
```
[77145.18920738 83053.58482997]
```

Due to `scikit-learn`'s focus on predictive modeling, it does not provide as much information about the model as `statsmodels`. There are, for example, no *p*-values or confidence intervals for the coefficients, as these are not as relevant in predictive modeling. In contrast, `scikit-learn` implements a variety of ways to validate models and estimate their performance. In the next section, we will see how to use `scikit-learn` to split a dataset into a training and a holdout set and estimate the performance of a model using the holdout set.

By default, `scikit-learn` will fit a model with an intercept. If you want to fit a model without an intercept, you can set the `fit_intercept` argument to `False` when you define the model.

```
model = LinearRegression(fit_intercept=False)
model.fit(X, y)
print(model.intercept_)
print(model.coef_)
```

Output
```
0.0
[33.51785714]
```

The intercept is now fixed at 0, and the coefficient is adjusted accordingly.

10.5.4 Splitting Datasets and Evaluating Model Performance

In order to split a dataset into training and holdout sets, we can use the *train_test_split* function from the `sklearn.model_selection` module.

```
from sklearn.model_selection import train_test_split

baseball = pd.read_csv("baseball_payroll.csv")
X = baseball[["Average Payroll (Million)"]]
y = baseball["Total Wins"]

X_train, X_holdout, y_train, y_holdout = train_test_split(X, y, ①
    test_size=0.2, ②
    random_state=123) ③
```

```
model_full = LinearRegression() ④
model_full.fit(X, y)
model_train = LinearRegression()
model_train.fit(X_train, y_train)

print(f"Full model intercept: {model_full.intercept_}")
print(f"Full model coefficients: {model_full.coef_}")
print(f"Model intercept: {model_train.intercept_}")
print(f"Model coefficients: {model_train.coef_}")
```

① The *train_test_split* function takes the independent (X) and dependent (y) variables as arguments and returns training and holdout sets for the independent and dependent variables.
② The `test_size` argument specifies the proportion of the dataset that should be used as the holdout set. Here, we split the dataset into 80% training and 20% holdout sets.
③ The `random_state` argument is used to set the random seed for reproducibility. You can also pass in a `numpy` random number generator.
④ We train two models. The first model is trained on the full dataset, and the second model is trained on the training set.

Output

```
Full model intercept: 207.47931041732843
Full model coefficients: [0.3904698]
Model intercept: 213.31899006194885
Model coefficients: [0.32644877]
```

The two models are similar but not identical. This becomes more obvious when we compare the two models by plotting them.

```
x_range = pd.DataFrame({'Average Payroll (Million)': [0, 250]})
fig, ax = plt.subplots(figsize=[6, 4])
ax.plot(x_range, model_full.predict(x_range), color='grey')
ax.scatter(X_train, y_train, color='grey')
ax.plot(x_range, model_train.predict(x_range), color='black',
   linestyle='dashed')
ax.scatter(X_holdout, y_holdout, color='black', marker='x')
ax.set_xlabel("Average Payroll (Million)")
ax.set_ylabel("Total Wins")
```

Figure 10.15 shows the comparison. The dashed line is the regression line of the model fitted using only the training set.

We now estimate the model performance by calculating the RMSE value.

```
from sklearn.metrics import mean_squared_error ①
rmse_full = mean_squared_error(y, model_full.predict(X),
   squared=False) ②
rmse_train = mean_squared_error(y_train,
   model_train.predict(X_train), squared=False)
```

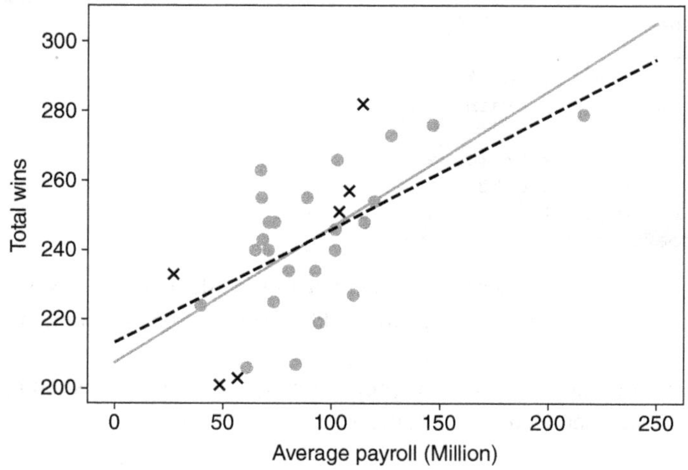

Figure 10.15 Linear regression model fitted to the full dataset set (gray line) and the training set (full circles and dashed line). The holdout set is shown as black crosses.

```
rmse_holdout = mean_squared_error(y_holdout,
   model_train.predict(X_holdout), squared=False)
print(f"Full RMSE: {rmse_full:.3f}")
print(f"Training RMSE: {rmse_train:.3f}")
print(f"Holdout RMSE: {rmse_holdout:.3f}")
```

① The `sklearn.metrics` module contains a wide variety of performance metrics for regression, classification, and more. Here, we import the *mean_squared_error* function, which by default calculates the MSE value, the square of the RMSE.

② The `squared=False` argument is used to get the RMSE value instead of the MSE value.

Output

```
Full RMSE: 16.904
Training RMSE: 15.701
Holdout RMSE: 21.643
```

The training set RMSE is considerably smaller than the holdout set RMSE. This is expected because the model was trained on the training set. Estimating the performance using data that were not included in the training, is more relevant and we should look at the holdout set RMSE to make our final assessment of the model performance.

Exercises

10.1 An online retailer wants to learn more about the success of its online advertising program. It reviews three weeks' worth of sales of a variety of different products at different price ranges and analyzes the data in a regression. The variables involved are sales (Y) and ad costs (X) (each in $000). The following information is obtained:

```
intercept: coef. = 32    SE = 5.6
X1: coef. = 1.6    SE = 1.9
```

a) Write out the regression formula that this output represents and interpret it.
b) What does "SE" stand for? What information does "SE = 1.9" give you, in a general sense?
c) For advertising expenditures of $10,000, what level of sales is predicted?
d) Around this prediction, use the SEs to place a range of sales in which you would have substantial confidence that actual sales will fall (multiply the X1 coef. and the intercept by +2 and −2 SEs).

10.2 A retail company wants to find out whether clickthroughs (CTs) are a good substitute for sales in evaluating the effectiveness of an online ad. One CT is one person clicking on an ad to learn more. Advertisers typically pay for ads on a per-clickthrough basis, so they want to learn quickly whether the ad is paying off. CTs have the advantage of being much more plentiful than sales, and accumulating faster, allowing the firm to judge quickly whether an ad is effective. Data on sales and CTs for 13 ads are recorded in *clickthroughs.csv*.
a) Create a scatterplot of the data. Do you see a linear relationship?
b) Calculate the correlation coefficient and assess whether CTs are a good substitute for sales.
c) Assume now that the company has determined that CTs are, in fact, a good proxy (substitute) for sales. Use regression to find a linear relationship between the two. Add the regression line to the scatterplot.
d) What level of sales would you predict for 350 CTs?

10.3 To facilitate sales of new cars, a large European auto retailer accepts trade-ins of used cars, which it must then resell. It wants to use a model to determine the resale value of value used cars based on their mileage (distance traveled).

a) Use the data in *toyota-km.csv* to develop a linear regression model that predicts sale price based on kilometers (KM) driven. Report the equation for the model, and interpret the coefficient for KM.
b) Predict the sale price for a vehicle that has 45,000 KM.

10.4 Large realtor agencies and online real estate companies like Zillow deal with large numbers of homes, both on and off the market, and need a quick way to estimate their value. One determinant of value is the size of the home.

a) Use the data in *WestRoxbury.csv* to fit a linear regression model to predict home value, with living area as the predictor.
   Hint: Use e.g. `Q("LIVING AREA")` to refer to the column names that have spaces in the formula.
b) What is the RMSE for this model?
c) Create a residual plot for the model.
   Hint: Use `alpha` to make the points semitransparent.
d) Create a scatterplot of the data with the regression line added.
e) Predict the value of a home that has 1800 sq ft. of living area.

11

Multiple Linear Regression

To this point, we have been concerned with only one predictor variable—salary in the baseball example, and cotton dust exposure in the pulmonary function example. Let us now turn to the more common situation where there are multiple independent or predictor variables. There are a variety of statistical and machine learning techniques that can model relationships between predictor variables and an outcome variable; we will focus on the oldest: multiple linear regression.

After completing this chapter, you should be able to:

- Fit a multiple linear regression model
- Test the statistical significance of coefficients
- Establish confidence limits around coefficients
- Distinguish traditional explanatory purposes from predictive modeling purposes, and identify the model performance metrics appropriate in each case
- Explain the role of training and holdout datasets
- Implement a predictive regression model and evaluate it using holdout data

11.1 Terminology

First, to minimize confusion, we will pause to review terminology. Different disciplines (statistics, computer science, IT) use different terms to refer to the variables in a regression. Figure 11.1 gives a summary.

There can also be different terms for the unit of observation, whether it is a patient in a health study, customer, web visitor, insurance claim, tax return, or whatever. All these terms, referring to that unit of observation, mean essentially the same thing:

- Subject
- Case

Statistics for Data Science and Analytics, First Edition. Peter C. Bruce, Peter Gedeck, and Janet Dobbins.

Companion website: www.wiley.com/go/Wiley_Statistics_for_Data

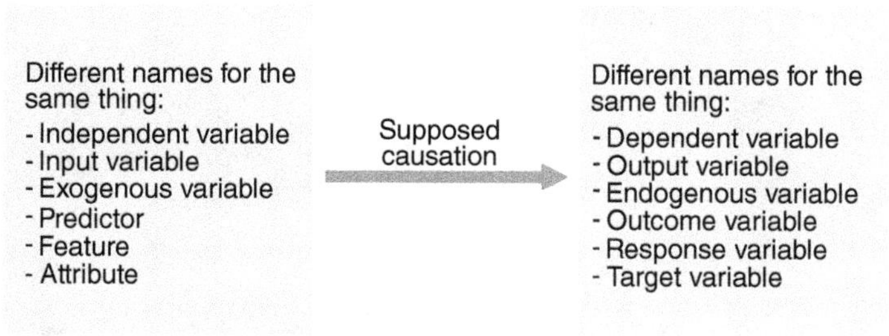

Figure 11.1 Different terms for variables in regression.

- Observation
- Example
- Sample

Note that the use of the term "sample" here, which comes from the computer science and machine learning community, means *one* observation, in contrast to the standard statistics usage where it refers to a collection of observations.

In the basic regression scenario, we have measurement data on an outcome variable—a single dependent variable of interest. We wish to model how that variable depends on other variables—input or independent variables—for which we also have data for the same observations.

The diagram 11.1 talks of "causation" and the terminology refers to one outcome variable "depending" on other variables. The directional nature of this relationship is a product of our belief, presumably on the basis of theory or knowledge, but regression does not prove it. The mathematics of regression merely describes a relationship; it does not prove a direction of causation. So the logical train of thought is thus: (1) We have a theory that y depends on a set of x variables. (2) Regression analysis may confirm that there is a relationship, and it may also describe the strength of that relationship. (3) If so, we take this as evidence that our theory is correct. However, you can see that there is no guarantee that the theory that y depends on x is correct. The direction of the relationship could be the reverse. Or both x and y could depend on some third variable.

11.2 Example—Housing Prices

The years 2002 to 2005 marked a housing bubble around the world, particularly in the United States. In 2006, prices started declining. In 2007, the decline accelerated, culminating in 2008 in a financial panic, massive intervention in financial markets by the US Federal Reserve and the start of a protracted and severe recession.

In retrospect, the growth of the bubble is clear. At the time, however, key actors failed to see it or ignored the signs. Before the bubble collapsed, many people concluded that their house was fairly priced if all other similar houses nearby were similarly priced.

Is it possible to establish an objective standard for home valuation, so that lenders and real estate brokers have a reference point more solid than the other houses comprising the bubble?

11.2.1 Explaining Home Prices

One possibility is to consider the underlying determinants of home value, and then establish a relationship between those variables and home prices. Until now, we have considered a single independent (predictor) variable and a single dependent (outcome) variable. It is unlikely that a phenomenon of interest—housing prices, in this case—can be explained by a single variable. Usually, multiple factors are at work, and we can model those in a multiple linear regression of the form

$$y = b_0 + b_1x_1 + b_2x_2 + \cdots + \varepsilon$$

Note this switch: When we introduced the equation for a line, and its slope and intercept, the intercept was denoted by "a." We now call the intercept b_0 in line with common usage for a multiple regression.

In words, this means that the dependent (outcome) variable y is a function of an unchanging value (a constant), b_0, plus a series of independent (predictor) variables x_1, x_2, etc., times their respective coefficients b_1, b_2, etc., plus an error term ε.

When we had just one independent variable, we saw that the chosen regression line minimized the sum of the squared deviations between the actual y values and the predicted y values that are points on the line. With multiple variables, the mathematics is more complex because we are no longer dealing with a line. However, the idea remains the same, which is to minimize squared deviations between predicted and actual values.

Notation

The dependent, i.e. outcome, variable is typically denoted by y.

Predicted values of y are typically indicated by the symbol $\hat{y}$ or "y-hat."

Independent variables in the abstract are usually referred to as "x variables"–$x_1, x_2, \ldots, x_i$.

The constant and the coefficients belonging to these x-variables are typically denoted by the letter b ($b_0, b_1, b_2, \ldots b_i$) or by the Greek letter beta ($\beta_0, \beta_1, \beta_2, \ldots, \beta_i$).

The error term, also called the residual, is usually denoted by the letter "e" or the Greek letter epsilon–ε.

11.2.2 House Prices in Boston

Let's consider a simplified example—the Boston Housing data from the 1970 census. (The complete data are available at the University of California at Irvine (UCI) Machine Learning Repository.) A reduced set of variables is used here. The outcome variable is the average price of homes by census tract, which is a neighborhood defined by the US Census Bureau. A tract consists of several thousand people and is relatively homogeneous with respect to demographic characteristics. The independent variables are factors that are believed to have an effect on home value. For simplicity, this presentation uses only a subset of the records and a subset of the variables originally associated with the data set (Table 11.1).

11.2.3 Explore the Data

As a first step, it would be useful to understand how these variables are distributed. We can do this with a set of separate boxplots, shown in Figure 11.2. Note that the variables do not share the same y-axis.

From the first boxplot, CRIM, we can see that most neighborhoods have very low crime rates with little variation among them and that there are some neighborhoods with much higher rates. You should take a moment to review the other boxplots to get a sense of how the variables they represent are distributed.

Next, it would be useful to learn the extent of correlation among the variables, especially the correlation between each independent variable and the dependent variable, which is median home value. We can do this with the correlation matrix shown in Table 11.2.

Definition: Correlation Matrix In a correlation matrix, the variable names constitute both the row and column headings. Each cell in the matrix indicates the correlation between its row and column variable.

For example, the correlation between RM, the number of rooms in a home, and MEDV, the median value of the home, is 0.36591, which is a low positive

Table 11.1 Boston Housing data variables.

Independent (predictor) variables	
CRIM	Crime rate per 1000 residents
RM	Average number of rooms per dwelling
Dependent (outcome) variable	
MEDV	Median home value in $1000s

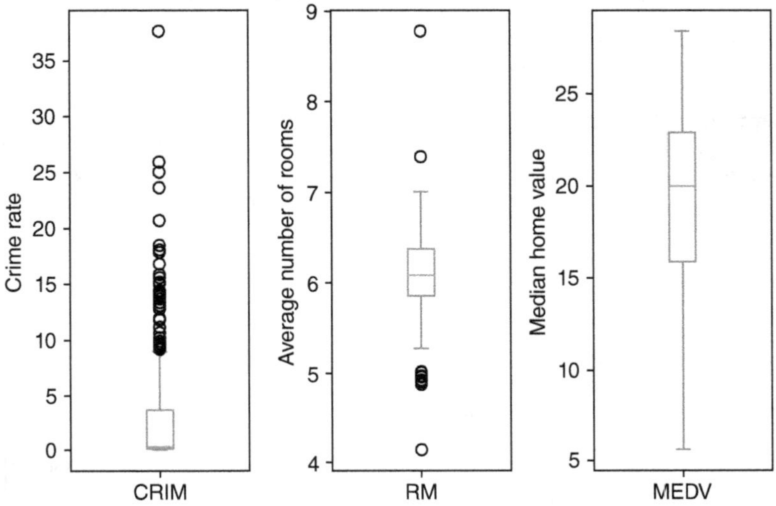

Figure 11.2 Boston Housing data variables.

Table 11.2 Boston Housing data: correlation matrix.

	CRIM	RM	MEDV
CRIM	1.00000	−0.11605	−0.60046
RM	−0.11605	1.00000	0.36591
MEDV	−0.60046	0.36591	1.00000

correlation. Also, the correlation between CRIM, the crime rate, and MEDV is −0.60046. As the crime rate decreases, the median value of the home generally increases. Along the diagonal is a series of 1's; the correlation of a variable with itself is 1. The cells above the diagonal are duplicative of the cells below the diagonal (the correlation of RM with CRIM is the same as the correlation of CRIM with RM).

11.2.3.1 Performing and Interpreting a Regression Analysis

Finally, we can perform a multiple regression analysis on the data using `statsmodels`. We specify MEDV as the dependent (outcome) variable and CRIM and RM as the predictor (independent) variables. Figure 11.3 shows the output.

Let's focus now on the coefficients and their interpretations. Ignore the other part of the output for now.

```
Results: Ordinary least squares
==================================================================
Model:              OLS              Adj. R-squared:     0.445
Dependent Variable: MEDV             AIC:                1431.9896
Date:               2023-05-27 11:06 BIC:                1442.7514
No. Observations:   267              Log-Likelihood:     -712.99
Df Model:           2                F-statistic:        107.8
Df Residuals:       264              Prob (F-statistic): 6.03e-35
R-squared:          0.449            Scale:              12.357
------------------------------------------------------------------
               Coef.   Std.Err.     t      P>|t|   [0.025  0.975]
------------------------------------------------------------------
Intercept      2.5088    2.8215    0.8892  0.3747  -3.0467  8.0643
CRIM          -0.4904    0.0399  -12.3025  0.0000  -0.5689 -0.4119
RM             3.0083    0.4606    6.5311  0.0000   2.1013  3.9152
------------------------------------------------------------------
Omnibus:              7.375        Durbin-Watson:          1.883
Prob(Omnibus):        0.025        Jarque-Bera (JB):       12.275
Skew:                 0.044        Prob(JB):               0.002
Kurtosis:             4.047        Condition No.:          101
==================================================================
Notes:
[1] Standard Errors assume that the covariance matrix of the
errors is correctly specified.
```

Figure 11.3 Multiple linear regression: Boston Housing data.

The CRIM coefficient is –0.4903722, and the RM coefficient is 3.0082633. The *y*-intercept is 2.5087717, but this is not a very meaningful number by itself in this case. (The intercept contributes to a better fitting regression within the range of the data, but the implied value of a house when all predictors are zero is a concept that doesn't make sense.)

We will round these coefficient values to four significant figures for brevity.

11.2.4 Using the Regression Equation

The regression equation is:

$$\text{MEDV}_r = 2.509 - 0.4904 \times \text{CRIM} + 3.009 \times \text{RM}$$

What would you predict for the following housing tract:

Crime rate: 2 (per 1000), Rooms: 8

We plug these values into the regression equation:

$$\begin{aligned}\text{MEDV} &= 2.509 - 0.4904 \times 2 + 3.009 \times 8 \\ &= 2.509 - 0.98 + 24.7 = 25.6\end{aligned}$$

So the predicted median value for this tract is $25,600 (recall that these data are from mid-20th century).

11.3 Interaction

The regression model that we have considered to this point tells a story in which the effect of each predictor variable is constant throughout the data. This is not always true.

Consider grapefruit and the drug quinidine. Alternative medicine practitioners advocate consuming grapefruit to minimize the symptoms of malaria. Quinidine is also prescribed by traditional doctors for malaria. So a study of the two variables might show that each, by itself, is associated with diminished malarial symptoms. However, there is an interaction between grapefruit and quinidine. Grapefruit affects certain stomach enzymes in ways that inhibit the absorption of quinidine. Therefore, if you take quinidine, adding grapefruit will probably exacerbate malarial symptoms, rather than adding further relief.

Let's now consider how we incorporate an interaction term into a regression model. Without interaction, a regression equation with two independent variables x_1 and x_2 and the random error term "ε," the Greek letter epsilon, looks like this.

$$y = a + b_1x_1 + b_2x_2 + \varepsilon$$

With an interaction term added, the equation becomes

$$y = a + b_1x_1 + b_2x_2 + b_3x_1x_2 + \varepsilon$$

You create the derived variable x_1x_2 by multiplying the two variables together.

Definition: Derived Variable A derived variable is a variable that is created from two or more other variables. A derived variable can be included in a regression model and treated like any other variable.

Let's look again at the Boston Housing data and create a new derived variable from CRIM, the crime rate, and RM, the number of rooms: CRIM*RM. The new regression output is shown in Figure 11.4.

This is the resulting regression equation, rounded to four significant figures.

$$\text{MEDV}_i = -7.624 + 1.247 \times \text{CRIM} + 4.680 \times \text{RM} - 0.2916 \times \text{CRIM} \times \text{RM}$$

Compared to the initial regression model without the interaction term, the effect of CRIM is now positive! Although the value is small, a higher crime rate predicts a higher MEDV.

```
Results: Ordinary least squares
====================================================================
Model:              OLS              Adj. R-squared:     0.485
Dependent Variable: MEDV             AIC:                1413.2655
Date:               2023-05-27 11:09 BIC:                1427.6145
No. Observations:   267              Log-Likelihood:     -702.63
Df Model:           3                F-statistic:        84.43
Df Residuals:       263              Prob (F-statistic): 2.74e-38
R-squared:          0.491            Scale:              11.477
--------------------------------------------------------------------
               Coef.  Std.Err.    t     P>|t|    [0.025   0.975]
--------------------------------------------------------------------
Intercept     -7.6236   3.4973 -2.1799 0.0302 -14.5098 -0.7374
CRIM           1.2469   0.3790  3.2898 0.0011   0.5006  1.9932
RM             4.6799   0.5733  8.1627 0.0000   3.5510  5.8089
CRIM:RM       -0.2916   0.0633 -4.6073 0.0000  -0.4162 -0.1670
--------------------------------------------------------------------
Omnibus:              16.819      Durbin-Watson:          1.860
Prob(Omnibus):        0.000       Jarque-Bera (JB):       38.218
Skew:                 0.251       Prob(JB):               0.000
Kurtosis:             4.784       Condition No.:          650
====================================================================
Notes:
[1] Standard Errors assume that the covariance matrix of the
errors is correctly specified.
```

Figure 11.4 Regression output with CRIM*RM interaction.

RM is slightly larger (more rooms increases value). The interaction term is significant—it has a very low *p*-value—and its effect is negative.

Let's look at the effect of increasing CRIM by one unit.

11.3.1 Original Regression with No Interaction Term

For the original regression, when CRIM is increased by one unit, the response changes as follows.

Original equation:

$$\text{MEDV}_r = 2.509 - 0.4904 \times \text{CRIM} + 3.009 \times \text{RM} \tag{11.1}$$

Same equation, but CRIM increased by 1 unit:

$$\begin{aligned}\text{MEDV} &= 2.509 - 0.4904 \times (\text{CRIM} + 1) + 3.009 \times \text{RM} \\ &= 2.509 - 0.4904 \times \text{CRIM} - 0.4904 + 3.009 \times \text{RM} \\ &= 2.0186 - 0.4904 \times \text{CRIM} + 3.009 \times \text{RM} \end{aligned} \tag{11.2}$$

The effect of increasing CRIM by 1 unit is:

$$\text{Eqs. (11.2)–(11.1)} = -0.4904$$

So, if we hold the number of rooms constant with no interaction term, then increasing the crime rate one unit decreases the average value of homes by $490.

11.3.2 The Regression with an Interaction Term

For the regression with an interaction term, the equation is as follows.

$$\text{MEDV}_i = -7.624 + 1.247 \times \text{CRIM} + 4.680 \times \text{RM} - 0.2916 \times \text{CRIM} \times \text{RM} \tag{11.3}$$

Now the same equation, but CRIM increased by 1 unit:

$$\begin{aligned}\text{MEDV} &= -7.624 + 1.247 \times (\text{CRIM} + 1) + 4.680 \times \text{RM} \\ &\quad - 0.2916 \times (\text{CRIM} + 1) \times \text{RM} \\ &= -7.624 + 1.247 \times \text{CRIM} + 1.247 + 4.680 \times \text{RM} \\ &\quad - 0.2916 \times \text{CRIM} \times \text{RM} - 0.2916 \times \text{RM} \\ &= -6.377 + 1.247 \times \text{CRIM} + 4.3884 \times \text{RM} - 0.2916 \times \text{CRIM} \times \text{RM}\end{aligned} \tag{11.4}$$

The effect of increasing CRIM by 1 unit is:

$$\text{Eqs. (11.4)–(11.3)} = 1.247 - 0.2916 \times \text{RM}$$

For RM = 6, an average sized house, this works out to −0.5026—holding number of rooms constant, adding a unit to the crime rate subtracts $502 from MEDV. The more rooms, the greater the negative impact from crime.

11.3.3 Does Crime Pay?

Let's return to the coefficient for CRIM: 1.247. It looks like higher crime rates produce higher home values, which makes no sense. But we just found that increasing CRIM actually *lowered* MEDV. We must look at the whole picture, taking account of the interaction, and increasing CRIM has an effect not just via the coefficient for CRIM, but also the coefficient for CRIM × RM. term.

Try It Yourself

What is the effect of increasing CRIM by one unit when RM = 5? What about when RM = 7? Work through subtracting Eq. (11.3) from Eq. (11.4). Do higher crime rates really increase home values?

11.4 Regression Assumptions

Multiple linear regression models work best when certain assumptions about the data are met. These assumptions, as well as methods for checking them, are listed below.

Assumption 1: *The observations are independent.*
This is usually assured by design: random selection in surveys and random assignment in experiments. In an observational study, the analyst has little control over how the data were collected, but should be aware of ways in which data may not be independent (e.g. collected from the same location, or from the same batch in a process). The main problem with nonindependent data is that you have less information than your sample size may suggest.

Assumption 2: *The relationship being investigated is, indeed, linear.* There may be strong patterns in the data, but they don't show up in a linear regression.

Assumption 3: *The variance of y does not change as x changes.*
Assumptions two and three are both tested by plotting residuals against predicted values. The resulting scatterplot should ideally be rectangular and random along the vertical axis, although some vertical stripes might be more dense than others. Figure 11.5 illustrates such a distribution produced by generating random numbers.
The scatter diagram in Figure 11.6 suggests some basis for questioning whether assumptions two and three hold; residuals are more positive for homes of less than $13K MEDV.

Assumption 4: *For every value of x, the error values of y are Normally distributed.*
This assumption matters primarily for rigorous formula-based inference (p-values and confidence intervals). If the residuals lie along a straight diagonal in what is called a QQ plot, then we can assume approximate normality (see for example Figure 11.7). However, for many applications, including those in data science, a simple check of predictive accuracy (which does not require formal statistical inference) may suffice and these normality assumptions are not relevant. Also, resampling methods that are less sensitive to this assumption are increasingly used in place of formula-based inference.

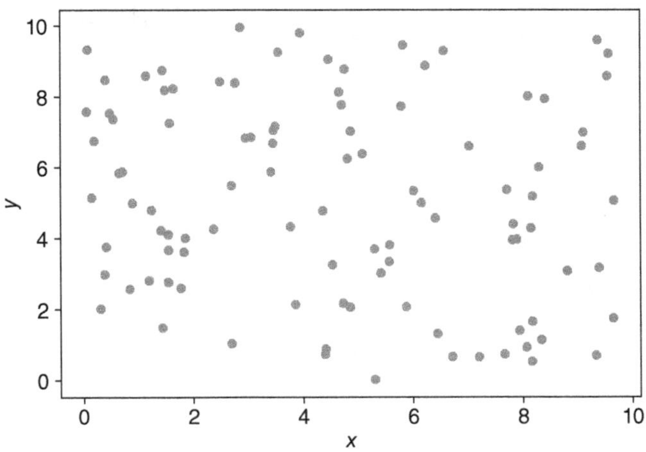

Figure 11.5 Random *x* and *y* coordinates, plotted.

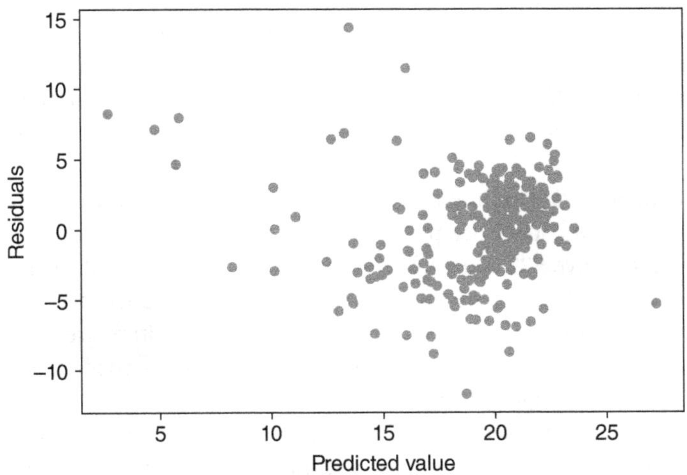

Figure 11.6 Boston Housing, predicted values vs. residuals.

11.4.1 Violation of Assumptions—Is the Model Useless?

If residuals are randomly and Normally distributed, this is evidence that the model is a good one—it has effectively accounted for all nonrandom sources of variation. Perfect models are hard to achieve, though. The worlds of natural phenomena and human behavior are complex and difficult to explain or predict completely. Usually, we must make do with models that fall short of 100% perfection—models where these assumptions may not be met.

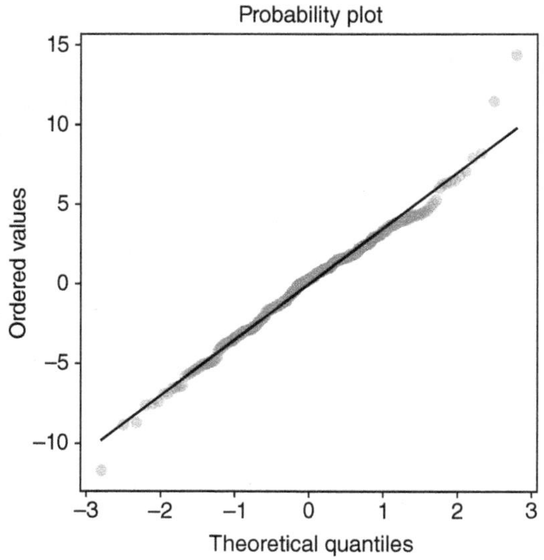

Figure 11.7 QQ-plot for residuals.

What happens if the assumptions are not met? Is the model useless? No. Violation of assumptions detracts from the credibility of the model as a full explanation of the relationship between the independent variables and the response variable, but it does not render the model valueless. Some knowledge is better than no knowledge.

In the housing data, for example, the residuals seem to violate assumptions at the extreme ends of the data. They are predominantly positive at predicted values below $15,000 and above $22,000. However, they are better behaved in the mid-range between these values. So, the model may be useful, after all, especially for non-extreme predicted values of MEDV. This is often the case with regression models—there may be a range over which the regression model is a good explanation of the data and other range or ranges over which it is less good.

Moreover, when regression is used for prediction rather than explanation, and a holdout sample is used to assess performance, the importance of these assumptions diminishes. In that case, our main concern is whether the model does a good job in predicting unknown values.

Now that we have learned how to fit a regression model to data, let's look at how well the model performs, keeping in mind whether our purpose is (1) regression for explanation and (2) regression for prediction.

11.5 Assessing Explanatory Regression Models

The standard regression evaluation metrics (such as those shown in Figure 11.4) computed in most statistical software, including `statsmodels` are for the data that the model was fit with. They show how well the model fits that data, but not necessarily how well the model will perform with new data.

11.5.1 Overall Model Strength R^2

R^2 is the measure of the extent of variation in y that is explained by the regression model. We already encountered it for the case of simple linear regression: R^2 is the square of the correlation coefficient ρ. An R^2 value of, for example, 0.75, means that the regression model explains 75% of the variation in y.

R^2 is a biased estimator. Its value in a sample will always be higher than the value you would get if you could perform a regression on the entire population. You can see this most easily in a small sample for a regression with multiple predictor variables. Consider the simplest case of all—two observations and two variables. The regression line will fit perfectly and R^2 will be 1.0, but such a model tells you next to nothing about the data.

`statsmodels` produces an adjusted R^2 that corrects for both the size of the sample and the number of predictor variables. The more predictor variables you have, the larger the sample required to get a useful regression estimate.

11.5.2 Assessing Individual Coefficients

The keys to using a regression model for explanatory purposes are the coefficients. These tell you *how much* a baseball team's record improves as the payroll grows, *how much* lung capacity diminishes as cotton dust exposure increases, *how much* consumer spending increases as advertising grows, etc. You will want to know, "Are the coefficient estimates reliable?"

11.5.3 Resampling Procedure to Test Statistical Significance

You can test the statistical significance of coefficients by repeatedly shuffling the outcome (dependent) variable and recalculating the regression each time. The algorithm (termed target shuffling by John Elder) is as follows:

1) Fit a regression model to the original data
2) Shuffle the outcome (dependent) variable
3) Recalculate the regression model and record the coefficients
4) Tabulate the distribution of these coefficients that the shuffled regression produced

Since the outcome variable was shuffled, any actual relationship between predictor variables and the outcome has been broken. The resampled predictor coefficients are the product of pure chance. If an observed coefficient value lies within a range which this chance shuffling produces, we can say that it might be the product of chance and is not statistically significant. If, on the other hand, an observed value rarely occurs in the distribution of its counterpart resampled values, we deem it statistically significant.

11.5.4 Resampling Procedure for a Confidence Interval (the Pulmonary Data)

Typically, coefficients in regressions are reported along with confidence intervals. This combination answers the question, "How differently might this estimate of the relationship turn out if we selected additional samples from the same population?" In most explanatory (research) situations, we do not have lots of additional samples to examine.

The next best thing is to take lots of bootstrap samples, i.e. resamples with replacement from our original sample. The steps are as follows.

Recall that the regression for the observed pulmonary suggested a negative relationship between pulmonary capacity and years of exposure data, with the regression equation:

$$y = -4.18458x + 424.5828$$

We want to establish a resampling confidence interval around both the constant and the slope coefficient. The algorithm is as follows:

1) Place N slips of paper in a box, where $N =$ the original sample size. On each slip of paper, write down the variable values for a single case. In this case, a pair for one case comprises the values for exposure and PEFR. A slip of paper for one such case, for example, might read (0,390) where zero is the exposure and 390 is the PEFR. If the data has multiple predictor variables, their values are also recorded on the slip of paper.
2) Shuffle the papers in the box, draw a slip, record its values, and replace it.
3) Repeat step 2 N times, yielding a bootstrap dataset of N records.
4) Perform a regression with PEFR as dependent variable and exposure as the independent variable, then record the coefficient and the constant.
5) Repeat steps 2–4, say, 1000 times.

You will end up with distributions of 1000 values for the coefficient (or coefficients if there are multiple predictors) and the constant, from which you can find appropriate percentiles to determine confidence intervals (e.g. the 5th and 95th percentiles for a 90% interval).

We saw earlier that the PEFR dataset has a lot of variability, and it is difficult to estimate where the regression line should be. We are definitely interested in the reliability of the estimated slope and y-intercept. We can use bootstrapping to assess this. Figure 11.8 compares the regression line of a single bootstrap sample to the regression line of the original data.

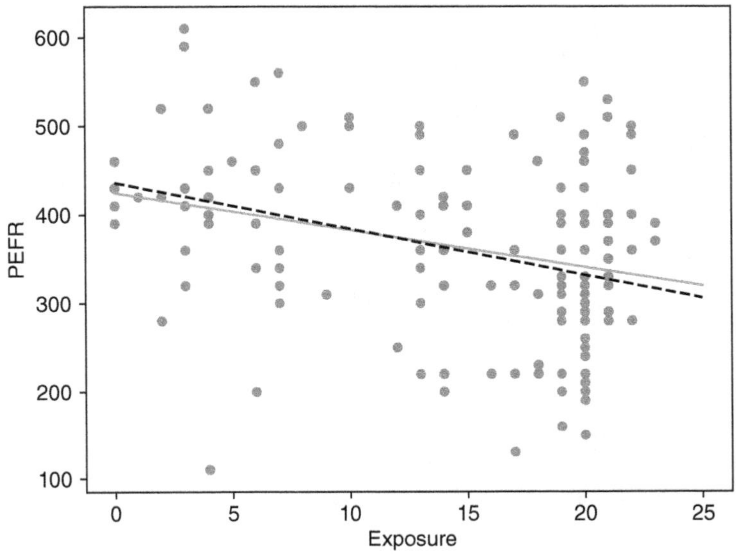

Figure 11.8 Regression via resampling—revisiting the PEFR data with a single bootstrap sample. The solid line is the regression line using the full dataset, the dashed line is fitted using a single bootstrap sample.

Producing a histogram for the resampled constant values will yield something like the output shown in Figure 11.9. The resampled intercept has a wide range of values. Also notice the 90% confidence interval (the 5th and 95th percentile values), running from around 390 to just under 460. You could easily choose a confidence degree other than 90%, we will see a 95% interval that is typical of software output in Section 11.5.5 (it will be wider, of course).

11.5.4.1 Interpretation

The regression equation for the observed data was

$$\text{PEFR} = -4.18458 * \text{EXPOSURE} + 424.5828.$$

The 90% resampling confidence interval for the constant (intercept) goes from 389 to 457.

> **Try It Yourself**
> Try the simulation again using the coefficient for exposure as the statistic of interest to obtain a resampling confidence interval for it.

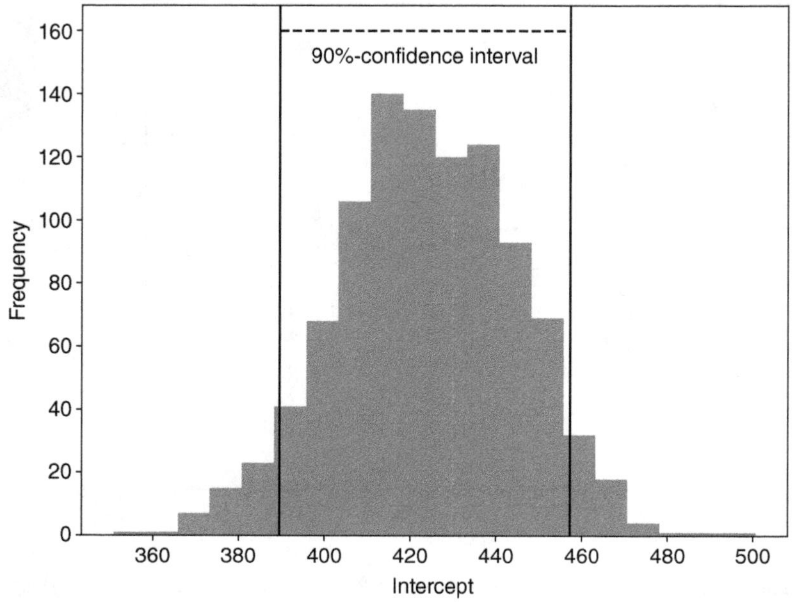

Figure 11.9 Analyzing PEFR regression intercept bootstrapped output.

11.5.5 Formula-based Inference

Before resampling was available, confidence intervals were calculated using formulas based on the *t*-statistic that we saw earlier. Over-simplifying a bit, and switching from the 90% interval we calculated via resampling to the 95% interval more typical of default software output, the interval is given by

$$b \pm t_{0.05}\mathrm{SE}$$

Where *b* is the coefficient from the regression, $t_{0.05}$ is the value of the *t* statistic at the 0.05 probability level, and SE is the standard error of *b*. The standard error is calculated as

$$\mathrm{SE} = \frac{s}{\sqrt{\sum x^2}}$$

Where *s* is the standard deviation of the residuals and *x* represents the individual values of the independent variable (to be squared and summed).

The `statsmodels` output, see for example Figure 11.10, shows 95% confidence intervals calculated automatically along the above lines.

11.5.6 Interpreting Software Output

The output from the `statsmodels` *summary2* method is daunting at first glance. In fact, you don't need to worry about most of it. For our purposes, let's focus on

Model output (`statsmodels`):

```
Results: Ordinary least squares
=================================================================
Model:                OLS               Adj. R-squared:     0.069
Dependent Variable: pefr                AIC:                1475.3501
Date:                 2023-05-27 11:51 BIC:                1480.9581
No. Observations:     122               Log-Likelihood:     -735.68
Df Model:             1                 F-statistic:        9.974
Df Residuals:         120               Prob (F-statistic): 0.00201
R-squared:            0.077             Scale:              10290.
-----------------------------------------------------------------
          Coef.    Std.Err.    t      P>|t|    [0.025    0.975]
-----------------------------------------------------------------
Intercept 424.5828  20.7960  20.4165  0.0000  383.4081  465.7575
exposure   -4.1846   1.3250  -3.1582  0.0020   -6.8079   -1.5612
-----------------------------------------------------------------
Omnibus:             0.767        Durbin-Watson:          1.111
Prob(Omnibus):       0.681        Jarque-Bera (JB):       0.891
Skew:                -0.162       Prob(JB):               0.641
Kurtosis:            2.734        Condition No.:          36
=================================================================
```

ANOVA analysis (`statsmodels`):

```
df          sum_sq        mean_sq           F      PR(>F)
exposure      1.0  1.026333e+05  102633.255269  9.974366  0.002008
Residual  120.0  1.234764e+06   10289.702381       NaN       NaN
```

Figure 11.10 Statsmodels regression output.

the coefficients section of the data found in the middle section. Notice the "Intercept" and "exposure" labels. Intercept refers to the y-intercept of the regression line, and exposure refers to the slope of the regression line. The Coefficients column (Coef.) supplies the values for the y-intercept and slope. Based on the data in Figure 11.10, with coefficients rounded to two decimals, the resulting equation for the regression line (rounded to two decimal places) is

$$y = -4.18x + 424.58$$

The standard error (Std.Err.) and t statistic are reported, as is a p-value (`P>|t|`) and the lower and upper bounds of a 95% confidence interval. The SE and t statistic are both used in calculating the p-value and the confidence interval, so are redundant information. The p-value is used in determining whether the coefficient value is different from 0 to a statistically significant degree. Confidence intervals and p-values are calculated by formula; for a better understanding of what they mean, review the resampling procedures presented above (Sections 11.5.3 and 11.5.4).

Note: In `statsmodels` you need to specify the degree of confidence for your interval—90%, 95%, etc. The default confidence interval is 95%. To change this to 90%, you can use the keyword argument `alpha=0.1` when calling the *summary* or *summary2* method. Figure 11.10 shows the output for the default 95% confidence level and interval.

Based on the data in Figure 11.10, the 95% confidence interval for the intercept, rounded to two decimal places, is 383.41 to 465.76. The 95% confidence interval for the slope is −6.81 to −1.56.

11.5.7 More Practice: Bootstrapping the Boston Housing Model

Again, we are interested in the coefficients and how reliable they are. If we took a different sample of housing tracts and performed a new regression, would we see completely different coefficients, or would they be similar?

The answer to that question, familiar by now, is given by drawing repeated bootstrap samples from our original sample, and repeating the regression over and over, recording the coefficients each time.

Here is the procedure applied to the Boston Housing data.

1) Place N slips of paper in a box, where $N =$ the original sample size. $N = 267$ for the *boston-housing.csv* dataset. On each slip of paper, write down the values for a single census tract. Here's how the slip of paper might look for one tract.

CRIM	RM	MEDV
8.98296	6.212	17.8

2) Shuffle the papers in the box, draw a slip, and replace.
3) Repeat step two N times.
4) Perform a regression with MEDV as dependent variable and the others as the independent variables. Record the coefficients.
5) Repeat steps two through four many times, say 1000.

You will end up with distributions of 1000 values for each coefficient. Figure 11.12 shows the distributions of the bootstrapped coefficients for the Boston Housing model. The distributions for the intercept and the coefficient for RM are both asymmetric which is also reflected in the calculated confidence intervals.

11.5.8 Inference for Regression—Hypothesis Tests

We have just finished looking at confidence intervals. Typical regression output will also include *p*-values for coefficients. Let's look again at the Boston Housing regression output, which is repeated in Figure 11.11.

```
Results: Ordinary least squares
=====================================================================
Model:              OLS                Adj. R-squared:      0.445
Dependent Variable: MEDV               AIC:                 1431.9896
Date:               2023-05-27 15:17   BIC:                 1442.7514
No. Observations:   267                Log-Likelihood:      -712.99
Df Model:           2                  F-statistic:         107.8
Df Residuals:       264                Prob (F-statistic):  6.03e-35
R-squared:          0.449              Scale:               12.357
---------------------------------------------------------------------
            Coef.    Std.Err.      t       P>|t|    [0.025    0.975]
---------------------------------------------------------------------
Intercept     2.5088    2.8215    0.8892   0.3747   -3.0467    8.0643
CRIM         -0.4904    0.0399  -12.3025   0.0000   -0.5689   -0.4119
RM            3.0083    0.4606    6.5311   0.0000    2.1013    3.9152
---------------------------------------------------------------------
Omnibus:              7.375        Durbin-Watson:           1.883
Prob(Omnibus):        0.025        Jarque-Bera (JB):        12.275
Skew:                 0.044        Prob(JB):                0.002
Kurtosis:             4.047        Condition No.:           101
=====================================================================
```

Figure 11.11 Multiple linear regression: Boston Housing data (same as Figure 11.3).

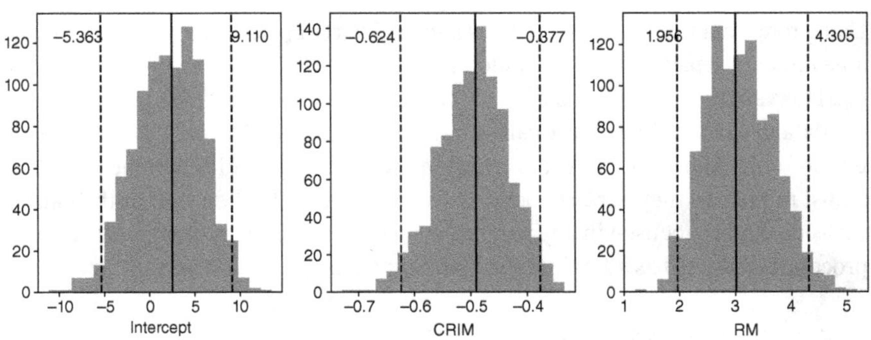

Figure 11.12 Bootstrapped coefficients for the Boston Housing model.

Consider the row for "RM."

- 3.008 is the estimated regression coefficient, which we covered earlier.
- 0.4606 is the standard error of the estimated coefficient RM. You can think of this as equivalent to the standard deviation of the bootstrapped coefficients that we produced above.
- 6.531 is the value of the *t*-statistic.
- < 0.0001 is the *p*-value.

The best way to understand the p-value is as the product of the resampling procedure outlined in Section 11.5.3.

The F-statistic reported in the Model row is the same F-statistic that we discussed earlier in the context of ANOVA. It indicates whether or not there is overall statistical significance considering all the variables.

11.6 Assessing Regression for Prediction

As noted above, standard regression metrics typically apply to the data used to fit the model. This makes some sense when our purpose is explanatory: to understand those specific data. However, it introduces optimistic bias when we seek to use the model to understand a broader population. The model will probably do better when we measure it with the same data it was fit to than when we apply it to new data.

When our purpose is prediction, the main issue is how well the model predicts new data. Prediction is primarily used in a big data setting, where data are plentiful, so we can afford to set aside some of the data where the outcome is known, for purposes of assessing the model.

11.6.1 Separate Training and Holdout Data

Using new data to evaluate a model is a formal part of predictive modeling. Available data are typically divided randomly into data used to fit the model and data used to evaluate it. The data used to fit the model are called the "training" data, and the data used to evaluate it are called, variously, "holdout," "validation," or "evaluation" data. The evaluation data are often used to compare different models, or to adjust models so they perform better. In some cases, a third portion of the data is set aside, and is not used in any part of the model fitting, adjustment and selection process, but is used as a final step in estimating true predictive performance.

11.6.2 Root Mean Squared Error—RMSE

The most popular metric to assess how well a regression model predicts new data is root mean squared error (RMSE). To calculate RMSE, you simply square the prediction errors (to render them all positive), find the mean, then take the square root (to put them back on the same scale as the original data.) RMSE is calculated on the holdout (validation) data.

There are a number of statistical and machine learning models besides linear regression that can be used for prediction; we will see another, k-nearest neighbors, in the next chapter. RMSE can be calculated for any model that predicts a numerical outcome, and so it is useful in comparing models against each other.

11.6.3 Tayko

Consider the hypothetical case of Tayko Software, a company that sells games and educational software.[1] It has recently put together a new catalog, which it is preparing to roll out in a direct mail campaign. In an effort to expand its customer base, it has joined a consortium of similar catalog firms. The consortium affords members the opportunity to mail catalogs to names drawn from a pooled list of customers totaling more than 20 million. Members supply their own customer lists to the pool and can borrow an equivalent number of names each quarter. A member can also apply predictive models to the entire consortium database to optimize its selection of names. Tayko has supplied 200,000 names from its own customer list. It is therefore entitled to mail to 200,000 names from the consortium per quarter. However, it has decided on a more limited effort and has budgeted funds for a mailing to 50,000 names. Although Tayko is a hypothetical company, the data in this case are from a real company that sells software through direct sales. The concept of a catalog consortium is also real, based on the Abacus Catalog Alliance.

The task is to build a regression model that will predict how much a customer will spend based on a small sample. Tayko will then apply this model to the consortium database to select the highest spending customers as its 50,000 names for mailing. These are the steps.

1) Draw a sample of names (including associated information for each name) from the consortium, mail catalogs to them, and see how much they purchase. A sample of 1000 should yield reasonably reliable results.
2) Randomly divide the 1000 names into a training partition and a validation partition (the validation sample is the holdout portion used for assessment).
3) Perform a regression with the training partition to determine the relationship between the available information for a customer and how much they spend.
4) Apply (score) that regression model to the validation partition and derive predicted spending values.
5) Use the actual and predicted spending values to calculate RMSE.
6) You can now try other regression models with different predictor variables to see how well they do. You can also try different types of models—see "*k*-nearest-neighbors" in the next section.

Deployment

1) Once you have selected the best model, it is time to apply that model to the entire consortium database of names to predict spending levels.
2) Select the 50,000 names with the highest predicted spending levels and mail

1 © Datastats LLC, used by permission.

Table 11.3 Tayko data fields.

source:_a	Was source company a? 1=yes, 0=no
source:_b	Was source company b? 1=yes, 0=no
source:_r	Was source company r? 1=yes, 0=no
Freq	How many orders in last 2 years
last_update_ days_ago	How many days ago was customer record last updated (inquiry or order)
Web order	Was order via web? 1=yes, 0=no

The customer information variables are shown in Table 11.3. "Source" refers to the catalog company that supplied the name to the consortium.

Before we proceed, however, we need to stop and consider how categorical variables can be included in regression equations.

11.6.4 Binary and Categorical Variables in Regression

Up to now we have used only continuous variables in regression. As you can see above, it is also possible to use binary—0 or 1—variables as independent or predictor variables, where a "1" or a "0" indicates a "yes" or "no" concerning the variable in question. The interpretation of the coefficient b in the resulting equation is simple. If the variable is present, e.g. "address is residential," the outcome is increased by an amount equal to b since it is multiplied by 1. If the variable is not present, e.g. "address is not residential," then the outcome is unaffected, since b is multiplied by 0 and the term reduces to 0.

Example

Suppose we want to model how much a catalog customer will spend in the coming year, and suppose the regression equation is

$$\text{Spending} = 65 + 35 \times \text{gender} + 0.0019 \times \text{income}$$

where

Gender: 1 if female, 0 if male

Income: Per capita income in the state of residence

You can see from the regression equation that being female increases predicted spending by \$35. Let's say a customer comes from a state where the per capita income is \$40,000.

$$\text{Spending if male: } 65 + 35 \times 0 + 0.0019 \times 40{,}000 = \$141$$

$$\text{Spending if female: } 65 + 35 \times 1 + 0.0019 \times 40{,}000 = \$176$$

Where there are more than two categories for a variable, such as a "source" that could be one of three different companies, you need to create multiple binary "dummy" variables. You can see this above, where source:_a, source:_b and source:_r are all listed as individual yes or no (0 or 1) variables. In other words, instead of a multi-category variable indicating "source could be a, b, or r," we have a series of binary variables indicating "is source = a?," "is source =b?," etc. Note that a binary variable is also a categorical variable, one with only two categories.

Definition: Dummy Variable A dummy variable is a derived variable created by taking the categories in a categorical variable and making separate binary (yes/no) variables for each of those categories.

11.6.5 Multicollinearity

Multicollinearity is another issue to be aware of when bringing multiple variables into the picture.

Definition: Multicollinearity Multicollinearity in multivariate modeling is the presence of one or more predictor variables that can be expressed as a linear combination of other predictor variable(s).

In this example, there are four possible sources, including "other." However, let's say that we had only two possible sources—"source a" and "source b." Every name comes from one source or the other. So the variable "source:_b" tells us nothing new—its information is exactly the same as the information contained in "source:_a." If a name comes from "source a," it cannot come from "source b," and vice versa.

Similarly, if we have four possible sources—"a," "b," "r," and "other"—and every name has one and only one source, then once we know the values for "a," "b," and "r," the value for "other" is predetermined. If a source is "a," "b," or "r," the value for "other" must be 0, and if a source is not "a," "b," or "r," the value for "other" must be 1.

In each case, we have multicollinearity: the information in one of the variables exactly duplicates the information contained in the other(s). This is a problem in regression, because the mathematical operations needed to calculate the regression crash in the presence of multicollinearity. Even if variables are only approximately, rather than exactly, duplicative, the regression calculations can be unstable and unreliable.

Multicollinearity can happen with both continuous and categorical—including binary—variables. To avoid multicollinearity, two checks are needed.

1) When creating binary dummy variables from a single categorical variable, use a maximum of $k - 1$ dummies, where k is the number of categories.
2) When making decisions about which variables to include in a regression, consider whether they might be measuring the same thing and check the correlation between the two variables. Where correlation is high, you should use only one of them.

For calculation purposes, you will get the same results no matter which of the multicollinear variables you omit, so you should retain the variables that are most informative. For example, in the above problem it would make sense to retain the source variables that contain the specific information and omit "other."

11.6.6 Tayko—Building the Model

Let's return to our goal, which is to build a model that predicts how much an individual will spend from a catalog. Our first step is to build a regression model using the training data for which we have known spending outcomes. To avoid multicollinearity problems, we use only the three specific source categories, leaving out "other." Table 11.4 shows the first few rows of the data. The regression output is shown in Figure 11.13.

Table 11.4 Tayko data, spending known (top-10 rows).

source:_a	source:_b	source:_r	Freq	last_update_ days_ago	Web order	Spending
0	1	0	2	183	1	128
0	0	0	2	194	0	127
1	0	0	1	161	0	174
0	0	0	1	73	0	192
0	0	1	2	147	1	386
0	0	0	2	73	0	174
0	0	0	1	123	1	189
0	0	0	1	165	0	90
0	0	0	2	147	1	352
0	0	1	9	43	1	639

```
                 Results: Ordinary least squares
==========================================================================
Model:                 OLS                 Adj. R-squared:       0.461
Dependent Variable:    Spending            AIC:                  6546.4415
Date:                  2024-01-22 08:48    BIC:                  6575.9437
No. Observations:      500                 Log-Likelihood:       -3266.2
Df Model:              6                   F-statistic:          72.02
Df Residuals:          493                 Prob (F-statistic):   2.82e-64
R-squared:             0.467               Scale:                28032.
--------------------------------------------------------------------------
                        Coef.  Std.Err.     t    P>|t|    [0.025   0.975]
--------------------------------------------------------------------------
Intercept              93.0655  22.8179  4.0786 0.0001  48.2332 137.8978
source:_a              42.7556  20.5935  2.0762 0.0384   2.2938  83.2174
source:_b               1.5061  37.7112  0.0399 0.9682 -72.5884  75.6007
source:_r              56.2754  29.8218  1.8871 0.0597  -2.3181 114.8688
Freq                   81.2010   4.8176 16.8549 0.0000  71.7354  90.6667
last_update_days_ago   -0.6559   0.1521 -4.3115 0.0000  -0.9548  -0.3570
Q('Web order')         -2.4575  15.0377 -0.1634 0.8703 -32.0034  27.0884
--------------------------------------------------------------------------
Omnibus:               352.373          Durbin-Watson:           2.176
Prob(Omnibus):         0.000            Jarque-Bera (JB):        5542.869
Skew:                  2.891            Prob(JB):                0.000
Kurtosis:              18.252           Condition No.:           578
==========================================================================
Notes:
[1] Standard Errors assume that the covariance matrix of the
errors is correctly specified.
```

Figure 11.13 Tayko data, multiple linear regression output.

11.6.7 Reviewing the Output

The regression equation is

$$\text{Spending} = 93.07 + 42.76 \times \text{source:\_a} + 1.51 \times \text{source:\_b} + 56.28 \times \text{source:\_r} + 81.20 \times \text{freq} - 0.656 \times \text{last\_update\_days\_ago} - 2.46 \times \text{web\_order}$$

Make sure you understand how you can derive this equation by looking at the software output.

We can see that the source catalogs "a" and "r" add to the spending estimate, by about $43 and $56, respectively. Each additional order (freq) in the customer's history adds about $81 to spending. Each additional day since the last contact (update) cuts the spending estimate by $0.656. *P*-values are all low for these variables. Catalog "b" as source, and whether an order comes from the web, seem to make little difference (coefficients are small, and *p*-values quite high).

11.6.8 Scoring the Model to the Validation Partition

A brief review of the model coefficients and their significance is useful, but our real goal is to see how the model does in predicting spending when applied to new data. Our proxy for new data is the validation partition that we set aside at the beginning. We will pretend that we lack spending information for these data and predict spending for each customer. This is also termed "scoring" the model; in the machine learning community it is called "inference" (not to be confused with the statistical inference of *p*-values and confidence intervals). Then we will compare our predictions to the actuals.

The first few rows of the data to be predicted are shown in Table 11.5.

The first step is to fill in the unknown values with your predictions. How? Just apply the regression model that we have already developed:

$$\begin{aligned}\text{Spending} = {} & 93.07 + 42.76 \times \text{source\_a} + 1.51 \times \text{source\_b} \\ & + 56.28 \times \text{source\_r} + 81.20 \times \text{freq} - 0.656 \\ & \times \text{last\_update\_days\_ago} \\ & - 2.46 \times \text{web\_order}\end{aligned}$$

Table 11.5 Tayko data, validation partition.

row	source:_a	source:_b	source:_r	Freq	last_update_ days_ago	Web order	Spending
1	0	1	0	3	148	1	?
2	0	0	0	3	145	0	?
3	0	0	0	1	45	1	?
4	0	0	0	6	70	0	?
5	1	0	0	4	24	1	?
6	0	0	0	6	74	0	?
7	1	0	0	1	156	1	?
8	0	0	0	2	27	1	?
9	1	0	0	1	125	0	?
10	0	0	0	1	105	1	?

Table 11.6 Predictions from the regression.

row	Actual Spending	Predicted Spending
1	?	239
2	?	242
3	?	142
4	?	534
5	?	442
6	?	532
7	?	112
8	?	235
9	?	135
10	?	103

For the first row above, this works out as follows:

$$\begin{aligned}\text{Spending} = {} & 93.07 + 42.76 \times 0 + 1.51 \times 1 + 56.28 \times 0 \\ & + 81.20 \times 3 - 0.656 \times 148 \\ & - 2.46 \times 1 \approx \$238.6\end{aligned}$$

Thus, for the rows shown in Table 11.5, the predictions are shown in Table 11.6.

Next we can reveal the actual values for Spending that were hidden in Table 11.5. These are shown in Table 11.7 Then we compare predicted to actual and find the differences: (Actual – Predicted) in Table 11.8 (note that the Actual–Predicted values have been rounded to integers). And, finally, we average the squared differences and find the square root; this is the root mean squared error (RMSE), see Table 11.9. If you perform the calculations your results may vary slightly due to rounding.

Doing so for the remainder of the data, beyond the few rows shown in the illustration above, is left as an exercise.

Why square the differences?

11.6.9 The Naive Rule

Before we leave the subject, we'd like to know whether the model is useful at all! For example, does it do better than simply predicting that everyone will be average (also called the naïve rule)?

Table 11.7 Actual spending in hold-out data revealed.

row	Actual Spending	Predicted Spending
1	136	239
2	261	242
3	43	142
4	389	534
5	394	442
6	588	532
7	160	112
8	50	235
9	233	135
10	54	103

Table 11.8 Actual spending in hold-out data revealed, adding residuals.

row	Actual Spending	Predicted Spending	Residual
1	136	239	−103
2	261	242	19
3	43	142	−99
4	389	534	−145
5	394	442	−48
6	588	532	56
7	160	112	48
8	50	235	−185
9	233	135	98
10	54	103	−49

The naive rule is simply predicting that everyone will spend at the average level in the training data (213). This rule, when applied to the few rows of the validation data shown above (see Table 11.10), has a RMSE of 172, considerably higher than the 98 in the regression model. This suggests the model does have some predictive power. In similar fashion we could compare the regression model to other models, most of which are beyond the scope of this book.

Table 11.9 Finding root mean squared error (RMSE).

row	Actual Spending	Predicted Spending	Residual	Squared residual
1	136	239	−103	10,609
2	261	242	19	361
3	43	142	−99	9801
4	389	534	−145	21,025
5	394	442	−48	2304
6	588	532	56	3136
7	160	112	48	2304
8	50	235	−185	34,225
9	233	135	98	9604
10	54	103	−49	2401
				Sum = 95,770
				Mean = 9577
				RMSE = 97.86

Table 11.10 Predicting everyone is average.

row	Actual Spending	Predicted Spending	Residual	Squared Residual
1	136	213	−77	5929
2	261	213	48	2304
3	43	213	−170	28,900
4	389	213	176	30,976
5	394	213	181	32,761
6	588	213	375	140,625
7	160	213	−53	2809
8	50	213	−163	26,569
9	233	213	20	400
10	54	213	−159	25,281
				Sum = 296,554
				Mean = 29,655
				Mean = 38,520
				RMSE = 172

11.7 Python: Multiple Linear Regression

11.7.1 Using Statsmodels

In Chapter 10, we learned how to make a linear regression with one independent variable. Extending this to multiple independent variables in `statsmodels` is straightforward. When we use the formula interface, we simply extend the formula with the additional independent variables. Let's look at the example from this chapter with the Boston Housing data (*housing.csv*). The dataset has two independent variables, crime rate `CRIM` and number of rooms `RM`, and one dependent variable, the median home value `MEDV`. When we write the formula, we use the + sign to add the independent variables on the right side. The formula for the multiple linear regression is therefore:

```
MEDV ~ CRIM + RM
```

That is essentially it. Fitting the model and getting the summary is the same as before.

```
import pandas as pd
import statsmodels.formula.api as smf

housing = pd.read_csv("boston-housing.csv")
formula = "MEDV ~ CRIM + RM"
model = smf.ols(formula, data=housing).fit()
print(model.summary2()) ①
```

Output

```
                Results: Ordinary least squares
==================================================================
Model:              OLS              Adj. R-squared:     0.445
Dependent Variable: MEDV             AIC:                1431.9896
Date:               2024-02-18 15:06 BIC:                1442.7514
No. Observations:   267              Log-Likelihood:     -712.99
Df Model:           2                F-statistic:        107.8
Df Residuals:       264              Prob (F-statistic): 6.03e-35
R-squared:          0.449            Scale:              12.357
------------------------------------------------------------------
               Coef.   Std.Err.     t      P>|t|    [0.025   0.975]
------------------------------------------------------------------
Intercept      2.5088    2.8215    0.8892  0.3747  -3.0467   8.0643
CRIM          -0.4904    0.0399  -12.3025  0.0000  -0.5689  -0.4119
RM             3.0083    0.4606    6.5311  0.0000   2.1013   3.9152
------------------------------------------------------------------
Omnibus:               7.375        Durbin-Watson:           1.883
Prob(Omnibus):         0.025        Jarque-Bera (JB):        12.275
```

```
Skew:                     0.044        Prob(JB):                  0.002
Kurtosis:                 4.047        Condition No.:             101
=====================================================================
Notes:
[1] Standard Errors assume that the covariance matrix of the
errors is correctly specified.
```

① The *summary2* method creates a slightly more compact view of the regression results, but contains otherwise the same information as the *summary* method.

The resulting model is

$$\text{MEDV}_r = 2.509 - 0.4904 \times \text{CRIM} + 3.009 \times \text{RM}.$$

Predicting new data is also unchanged. We use the *predict* method of the model object and call it with a dataframe that contains the new data.

```
new_data = pd.DataFrame({"CRIM": [2, 10], "RM": [8, 7]})
predictions = model.predict(new_data)
print(predictions)
```

Output

```
0    25.594134
1    18.662893
dtype: float64
```

11.7.1.1 Adding Interaction Terms

We can also define interactions in the formula. There are two ways of doing this:

```
MEDV ~ CRIM + RM + CRIM:RM
MEDV ~ CRIM * RM
```

Both of these formulas will create the interaction term between `CRIM` and `RM`. The first formula is more explicit, while the second is a shorthand that creates the interaction term (`CRIM:RM`) and the main effects (`CRIM` and `RM`).

Computationally, the interaction term is the product of the two independent variables. The model will then have the following form:

```
formula = "MEDV ~ CRIM + RM + CRIM:RM"
model_interaction = smf.ols(formula, data=housing).fit()
print(model_interaction.summary2())
```

Output

```
                Results: Ordinary least squares
=====================================================================
Model:                 OLS                 Adj. R-squared:     0.485
.....
R-squared:             0.491               Scale:              11.477
```

```
--------------------------------------------------------------------
                 Coef.   Std.Err.      t      P>|t|    [0.025   0.975]
--------------------------------------------------------------------
Intercept      -7.6236    3.4973   -2.1799  0.0302  -14.5098  -0.7374
CRIM            1.2469    0.3790    3.2898  0.0011    0.5006   1.9932
RM              4.6799    0.5733    8.1627  0.0000    3.5510   5.8089
CRIM:RM        -0.2916    0.0633   -4.6073  0.0000   -0.4162  -0.1670
--------------------------------------------------------------------
.....
```

The two models are fairly similar. The r^2 value is slightly higher for the model with the interaction term; 0.491 compared to 0.449. The same is true for the adjusted r^2 value, 0.485 compared to 0.445.

Figure 11.14 shows the actual vs. predicted values for the two models. The model with the interaction term has a slightly better fit, we can see that the points are closer to the line $y = x$.

```
fig, axes = plt.subplots(ncols=2, figsize=[8, 4])
df = pd.DataFrame({"Predicted value": model.fittedvalues,
                   "Actual value (MEDV)": housing["MEDV"]})
ax = df.plot.scatter(x="Actual value (MEDV)", y="Predicted value",
    alpha=0.5, ax=axes[0])
ax.set_title("(a) Main effects model")
ax.plot([0,30], [0,30], color="black")

df = pd.DataFrame({"Predicted value":model_interaction.fittedvalues,
                   "Actual value (MEDV)": housing["MEDV"]})
ax = df.plot.scatter(x="Actual value (MEDV)", y="Predicted value",
                     alpha=0.5, ax=axes[1])
ax.set_title("(b) Interactions model")
ax.plot([0,30], [0,30], color="black")
plt.show()
```

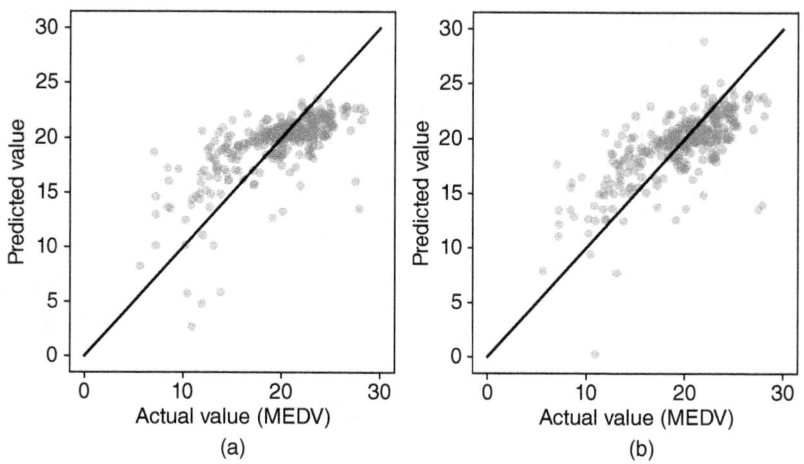

Figure 11.14 Actual vs. predicted MEDV plots for the main effects model (a) and the interactions model (b).

11.7.2 Diagnostic Plots

In Chapter 10, we learned about the residual plot. The same plot can also be created for multiple linear regression models. This chapter introduced the QQ-plot to analyze the distribution of the residuals and check for Normality. This is not so essential in most data science applications, but we show it here since you may encounter it. We can use the *qqplot* function from the `scipy` package to create the QQ-plot. Let's create both diagnostic plots for the interaction model. Figure 11.15 shows the residual plot and the QQ-plot.

```
import scipy.stats as stats

fig, axes = plt.subplots(ncols=2, figsize=[8, 4])
ax = axes[0]
ax.scatter(model_interaction.fittedvalues, model_interaction.resid,
   alpha=0.5)
ax.axhline(0, color="black", linestyle="dashed")
ax.set_xlabel("Fitted values")
ax.set_ylabel("Residuals")
ax.set_title("(a) Residual plot")

ax = axes[1]
stats.probplot(model_interaction.resid, dist="norm", plot=ax) ①
ax.get_lines()[0].set_color("C0") ②
ax.get_lines()[0].set_alpha(0.5)
ax.get_lines()[1].set_color("black")
ax.set_title("(a) QQ-plot")
plt.tight_layout()
plt.show()
```

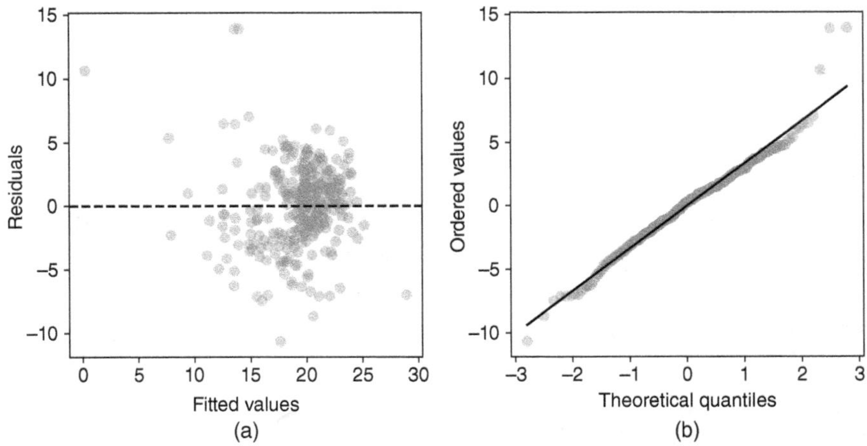

Figure 11.15 Residual plot (a) and QQ-plot (b) for the interaction model.

① The *probplot* function creates the QQ-plot. We use the `dist` argument to specify the expected theoretical distribution of the residuals. In this case, we use the Normal distribution. We didn't really need to specify this; we use this plot to check for Normality so the Normal distribution is the default.
② This and the following two statements modify the colors in the graph. We need to use this here because the *probplot* function doesn't expose this functionality.

11.7.3 Using Scikit-learn

In the last chapter, we provided the independent variable to the `LinearRegression` model as a dataframe with a single column. When we have multiple independent variables, we just add additional columns for each of the independent variables.

```
from sklearn.linear_model import LinearRegression
import pandas as pd

predictors = ["CRIM", "RM"] ①
outcome = "MEDV"
X = housing[predictors] ②
y = housing[outcome]
model = LinearRegression()
model.fit(X, y)
print(model.intercept_)
print(pd.Series(model.coef_, index=predictors))
```

① We define the independent variables in a list. This way, we can use the same list to more easily identify the coefficients.
② We use the list of predictors to extract a subset of the dataframe. X will have two columns.

Output

```
2.508771666893967
CRIM   -0.490372
RM      3.008263
dtype: float64
```

Predicting new data is also the same as before. We create a dataframe with the new data and call the *predict* method of the model.

```
new_data = pd.DataFrame({"CRIM": [2, 10], "RM": [8, 7]})
predictions = model.predict(new_data)
print(predictions)
```

Output

```
[25.59413421 18.66289324]
```

11.7.3.1 Adding Interaction Terms

We learned that adding an interaction term is equivalent to adding a column with the product of the two independent variables. Here is one way to do this.

```
X["CRIM:RM"] = X["CRIM"] * X["RM"]
model_interaction = LinearRegression()
model_interaction.fit(X, y)
print(model_interaction.intercept_)
print(pd.Series(model_interaction.coef_, index=predictors + ["CRIM:RM"]))
```

Output

```
-7.623594886934143
CRIM         1.246896
RM           4.679950
CRIM:RM     -0.291616
dtype: float64
```

We need to manage the names of the columns ourselves.

11.7.4 Resampling Procedures

In this section, we will implement the resampling procedures from Chapter 10. We start by using the shuffling (permutation) procedure to estimate statistical significance.

11.7.4.1 Estimating the Significance of the Coefficients

We first look at the statistical significance of the coefficients using again the *housing.csv* dataset. We use the `scikit-learn` library for this. For demonstration purposes, we will also add a random column to the dataframe.

```
from sklearn.linear_model import LinearRegression
import numpy as np

rng = np.random.default_rng(seed=321)

housing["RANDOM"] = rng.random(len(housing)) ①
predictors = ["CRIM", "RM", "RANDOM"]
outcome = "MEDV"
X = housing[predictors]
y = housing[outcome]

model = LinearRegression()
model.fit(X, y)
actual = pd.Series([model.intercept_, *model.coef_], ②
                   index=["Intercept", *predictors])

resamples = []
for _ in range(1000):
    model.fit(X, rng.permutation(y)) ③
```

```
    resamples.append((model.intercept_, *model.coef_))
resamples = pd.DataFrame(resamples, columns=["Intercept", *predic-
tors])
```

① We add a random column to the dataframe. This will be used for illustration purposes, to demonstrate that the random variable is not significant.
② We create a series with the actual coefficients. This allows us later to access the information by name.
③ The random number generator method *permutation*, returns a shuffled version of the outcome.

We can now create histograms of the coefficients and compare them to the actual coefficients.

```
fig, axes = plt.subplots(ncols=4, figsize=[10, 3])
for ax, name in zip(axes, resamples.columns):
    resamples[name].plot.hist(bins=30, ax=ax)
    ax.axvline(actual[name], color="black", linestyle="dashed")
    ax.set_xlabel(name)
plt.tight_layout()
plt.show()
```

Figure 11.16 shows the histograms. For the intercept and the two independent variables, `RM` and `CRIM`, the actual coefficients are well separated from the range of the resampled coefficients (i.e. the coefficients produced by fitting the model with a shuffled outcome variable). The random variable `RANDOM` on the other hand falls withing the distribution of resamples values. Unsurprisingly, the random variable is not significant in the model and can be removed.

11.7.4.2 Estimating Confidence Intervals—The Bootstrap

To estimate confidence intervals for the model coefficients, we create bootstrap samples of the data and fit the model to each sample. We then calculate the 2.5th and 97.5th percentiles of the coefficients to get a 95% resampling confidence interval. We will use the *housing.csv* dataset and the `scikit-learn` package for this.

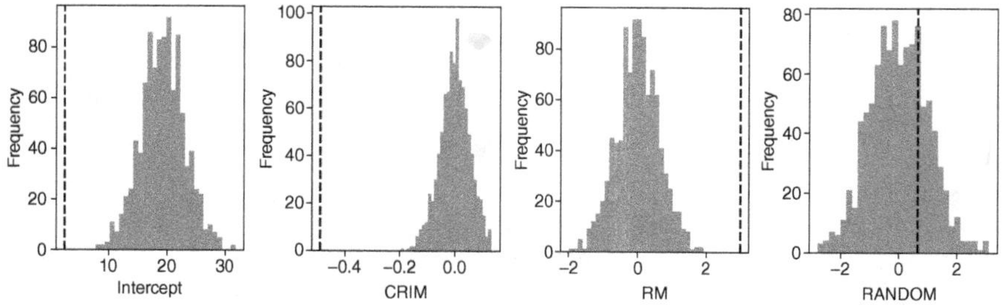

Figure 11.16 Histograms of the coefficients from the resampling procedure. The black vertical lines indicate the actual coefficients.

```
from numpy.random import RandomState
from sklearn.utils import resample

rng = np.random.RandomState(seed=321) ①
model = LinearRegression()
model.fit(X, y)
estimate = pd.Series([model.intercept_, *model.coef_],
                     index=["Intercept", *predictors]) ②
coefficients = []
for _ in range(1000):
    X_resampled, y_resampled = resample(X, y, random_state=rng) ③
    model = LinearRegression()
    model.fit(X_resampled, y_resampled)
    coefficients.append([model.intercept_, *model.coef_])
coefficients = pd.DataFrame(coefficients, columns=estimate.index)

conf_intervals = pd.DataFrame({
    "Coefficient": estimate, ④
    "Lower": np.percentile(coefficients, 2.5, axis=0),
    "Upper": np.percentile(coefficients, 97.5, axis=0)
})
print(conf_intervals)
```

① The *resample* function does not yet support the new `numpy` random number generator. We use the old `RandomState` class to create a random number generator.
② We store the estimates in a `pandas` Series. This allows us to easily access the coefficients by name.
③ We use the *resample* function from the `scikit-learn` package to create a bootstrap sample of the data. We then fit the model to the resampled data and store the coefficients in a list.
④ We calculate the 2.5th and 97.5th percentiles of the coefficients and store them in a dataframe.
⑤ The index of the series, i.e. the names of the estimated values, is used as the index of the dataframe. This way, the confidence intervals are easily matched to the coefficients.

Output

```
           Coefficient     Lower     Upper
Intercept     2.310490 -5.394596  8.763676
CRIM         -0.489569 -0.616600 -0.381332
RM            2.987768  1.972809  4.263446
RANDOM        0.646785 -0.819860  2.017485
```

Let's see how these compare to the confidence interval calculated by the `statsmodels` package using formulas (not resampling).

```
formula = "MEDV ~ CRIM + RM + RANDOM"
model = smf.ols(formula, data=housing).fit()
print(model.summary2())
```

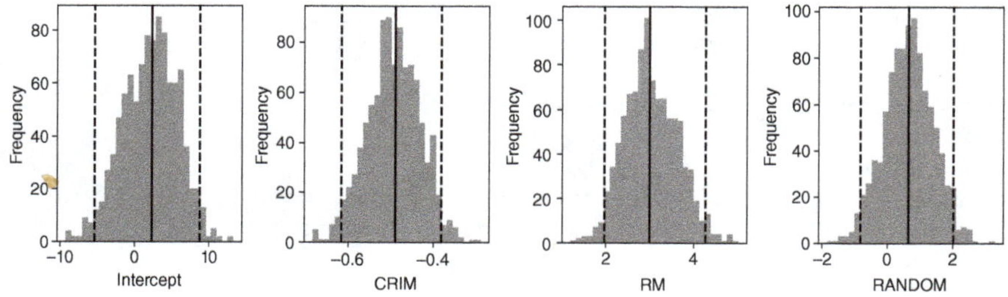

Figure 11.17 Distribution of resampled coefficients and the 95% confidence intervals (dashed lines). The black vertical lines shows the actual estimate of the coefficient.

Output

```
....
-----------------------------------------------------------------------
             Coef.   Std.Err.        t     P>|t|    [0.025    0.975]
-----------------------------------------------------------------------
Intercept   2.3105    2.8325    0.8157   0.4154   -3.2668    7.8878
CRIM       -0.4896    0.0399  -12.2726   0.0000   -0.5681   -0.4110
RM          2.9878    0.4615    6.4744   0.0000    2.0791    3.8964
RANDOM      0.6468    0.7581    0.8532   0.3943   -0.8459    2.1395
-----------------------------------------------------------------------
......
```

The confidence intervals are fairly similar. Figure 11.17 shows the distribution of the resampled coefficients and the 95% confidence intervals.

```
fig, axes = plt.subplots(ncols=4, figsize=[10, 3])
for ax, name in zip(axes, coefficients.columns):
    coefficients[name].plot.hist(bins=30, ax=ax)
    ax.axvline(actual[name], color="black")
    ax.axvline(conf_intervals.loc[name, "Lower"], color="black",
        linestyle="dashed")
    ax.axvline(conf_intervals.loc[name, "Upper"], color="black",
        linestyle="dashed")
    ax.set_xlabel(name)
plt.tight_layout()
plt.show()
```

Exercises

11.1 This problem uses a variant of the Boston Housing data (*boston-housing-large.csv*). This version of the data contains five predictor variables for MEDV. The five variables are:

- CRIM: crime rate per 1000 persons
- NOX: ambient nitric oxide concentration (parts per 10 million)

- RM: average number of rooms per dwelling
- PTRATIO: pupil teacher ratio by town
- LSTAT: percent of the population in a lower socioeconomic status

a) Make a scatterplot of MEDV vs. CRIM. What do you see? On the basis of your scatterplot, does CRIM appear helpful in predicting MEDV?

b) Run a regression of MEDV as a function of CRIM. Report a *p*-value and interpret it. From this, does CRIM appear useful in predicting MEDV?

c) Make a scatterplot of MEDV vs. LSTAT. What do you see? On the basis of your scatterplot, does LSTAT appear helpful in predicting MEDV?

d) Run a regression of MEDV as a function of LSTAT. Report a *p*-value and interpret it. From this, does LSTAT appeal helpful in predicting MEDV?

e) So far, we have been analyzing one predictor variable at a time, so we do not have a good idea how they work together to affect MEDV. Run a multiple regression of MEDV vs. all five of the predictor variables. Report the regression equation and the predictor variable *p*-values from the computer output. From this, does the regression appear helpful in predicting MEDV?

11.2 Use the dataset *Tayko-known.csv* for the following problems.

a) Perform a multiple linear regression. Specify appropriate independent (predictor) and response variables. Report the resulting equation.
Hint: This duplicates an illustration in the chapter.

b) Create a graph showing actual vs. predicted spending.

c) Calculate the following statistics for the *Tayko-known.csv* dataset and the regression.
   - standard deviation of the spending
   - mean spending
   - RMSE

   In a sentence or two describe variability in the *Tayko-known.csv* data and your regression. Your goal is to convey to your reader an understanding of average spending, how variable it is, and how much a typical prediction might be in error.

d) Use the regression equation to predict spending levels for the customer records in *Tayko-unknown.csv*.

e) Sort the results from the previous step by predicted spending, and report the top 10 customers for predicted spending.

f) Your goal is to generate at least $250 in spending per catalog mailed. How many customers should you mail to, from *Tayko-unknown.csv*?
Hint: Would you mail a catalog to the customer represented by the top row in the sorted results? The second row? The last row?

11.3 **Background**: Use the *trade-discount.csv* dataset for this problem. Consumer packaged goods (CPG) companies sell some goods online but generate most of their revenue from sales through brick and mortar retail companies. A CPG company typically manages a number of brands, each of which has multiple products. For example, Unilever, a large CPG based in the United Kingdom, owns over 400 brands. Just one of its brands, Dove, sells body washes, hand and body lotions, facial cleansers, deodorants, shampoos, conditioners and hair styling products. Each product is sold in multiple sizes and variations (e.g. different soap scents), so the number of individual "stockkeeping units" (SKU's) on offer from a CPG might be in the tens of thousands. A typical grocery or drug store has room to stock only a small fraction of all the products on offer from CPGs. A major challenge for the CPG is obtaining and retaining "shelf space" at major retailers. CPG companies promote their products through advertising, issuance of discount coupons to consumers, and "trade discounts" offered to retailers. Trade discounts are product-specific discounts offered to individual retailers that can fund co-advertising, or simply serve as an incentive to the retailer to promote the discounted product.

The Problem: You work for a CPG company that wants to evaluate its trade discount spending. The finance department, noting the high cost of the program, wants to curtail the discounts. The marketing and sales department, naturally, resists. It turns out that nobody has a good answer to the question, "Do discounts increase sales?" The company could try reducing the discounts and seeing if that made a difference in sales but that would entail substantial risk and upheaval. You suggest taking a look at existing data, to see if there is a relationship between trade discounts and sales revenue. (Note: There are problems with a similar scenario and similar data but different questions in Chapters 2 and 4)

a) You decide to perform a multiple linear regression, with Sales as the outcome variable, and trade discount percentage (Trade-disc) as a predictor variable. Why use trade discount percentage as a predictor rather than actual trade discount dollars?

b) You also decide to add Coupon as a predictor, so that the effects of trade discount spending and consumer coupon spending are kept separate. Perform a regression with Sales as the outcome variable and Trade-disc and Coupon as predictor variables. Report the resulting regression equation.

c) Consider an item where the trade discount is 0.26 (26%) and the coupon spending rate is 0.15 (15%). What is the predicted sales revenue?

d) Change the trade discount to 0.27 and recalculate predicted sales revenue. What is the increase in sales revenue?

e) This increased revenue has a cost—the larger trade discount the company must offer. How much is this cost in dollars?
f) What is the predicted net profit effect of increasing the trade discount?
g) Reviewing the regression statistics, comment on the reliability of the estimated relationship between sales, trade discount, and coupon spending.

12

Predicting Binary Outcomes

In this chapter we discuss a simple to understand technique for predicting binary outcomes—k-nearest-neighbors (k-NNs). After completing this chapter, you should be able to:

- Distinguish between predicting a numerical outcome and predicting a binary outcome
- Classify binary outcome data using k-nearest neighbors in Python
- Divide the data into training and validation partitions to optimize the choice of k
- Explain whether k-nearest-neighbors can be used for explanatory modeling

Many, if not most, statistical analyses and decisions involve binary outcomes: a consumer buys or doesn't buy, a prospect responds or doesn't respond, a patient dies or survives, a loan is paid off or goes into default, a person votes for your candidate or not, a link is clicked or not, etc. The structure of the data is the same as we have seen in multiple linear regression, except the outcome (dependent) variable is binary instead of continuous. Another term for this is classification—predicting which class a new record will fall into. In this case there are two classes.

Regression analysis is not suitable for modeling such data. In fact, despite the seeming simplicity of yes/no data, statistical methods for modeling it are actually more involved than those for numeric data and, therefore, are rarely touched upon in an introductory course.

One such method for prediction is intuitively simple, however, and merits a brief discussion.

12.1 *K*-Nearest-Neighbors

Recall that, in prediction, we "train" models with a set of data in which both the predictor and outcome variables are known. The goal is to predict an outcome

Statistics for Data Science and Analytics, First Edition. Peter C. Bruce, Peter Gedeck, and Janet Dobbins.

Companion website: www.wiley.com/go/Wiley_Statistics_for_Data

of interest for a new record, where we know only the values for the predictor variables. For example, a bank might have prior data on 5 year old loans for which a borrower's assets, debt, income, education and credit score are known, and the current status of the loan (current or in default) is also known. For new applicants, the bank can collect the same information on assets, debt, income, education and credit score, and, using the information from the prior loans, predict which class, current or in default, the loan will be in, five years later.

The k-nearest-neighbor method is quite simple. The basic idea is as follows:

1) Locate the k prior customers who are most like ("near") the new customer in terms of the predictor variables
2) Find which class most of those similar customers belong to
3) Predict that majority class for the new customer

Let's first illustrate this in the two-dimensional case where we wish to predict whether a household owns a riding mower or not. Figure 12.1 is a scatterplot illustrating our two predictor variables, the income and lot size of households that own riding mowers (filled circles), and households that do not (empty circles). It also shows a new household (the x) for which we want to predict whether a riding mower is owned.

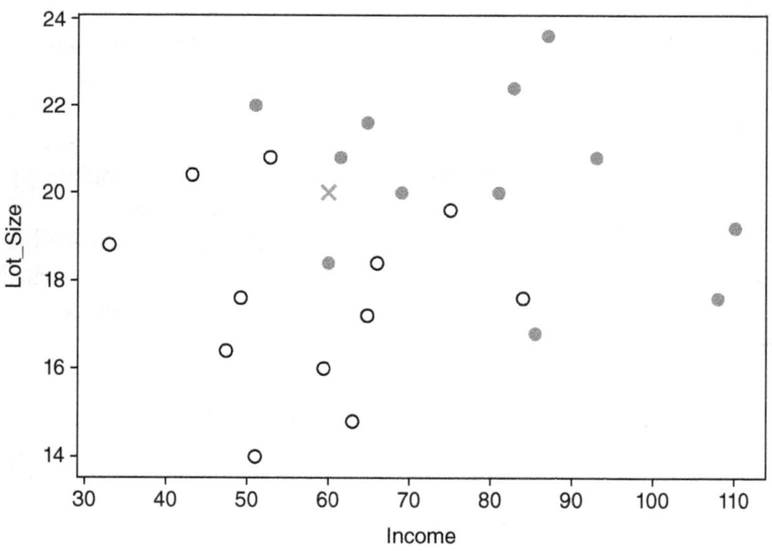

Figure 12.1 Riding Mower, classifying new household (cross) as owner (filled circle) or nonowner (open circle).

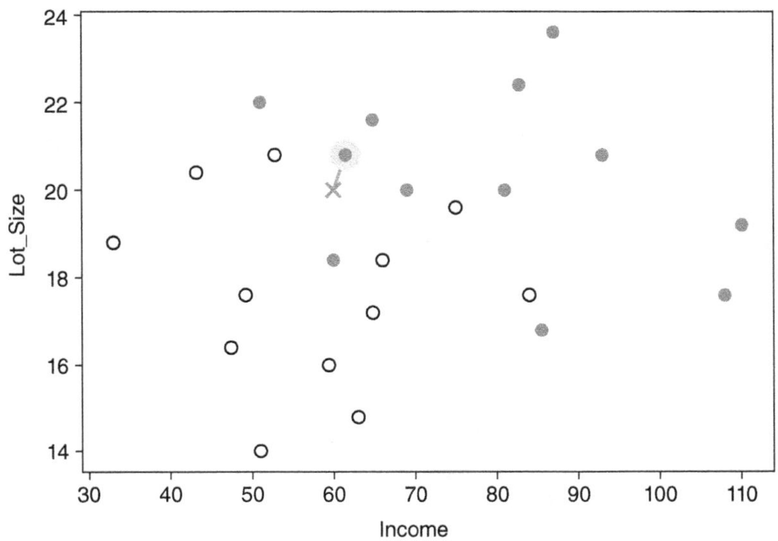

Figure 12.2 Finding the nearest single neighbor ($k = 1$).

Using the nearest-neighbor algorithm, we can locate the single closest household and determine its class (see Figure 12.2). In this case, it is "owner" so that is how we would classify our new household.

Let's expand the comparison base, so that we are considering the five nearest neighbors, not just the single nearest neighbor (Figure 12.3). Four of them are owners, one is a nonowner, so, by majority vote, we again classify the new household as an owner.

Determining the nearest neighbors can be done visually with just two variables, as above. To locate nearest neighbors in data with multiple predictor variables, we use the concept of distance that we introduced in Section 3.4. We illustrate this in the case of the big retailer Target, introduced in Chapter 1, and how it used statistical methods to predict whether a customer was pregnant or not.

12.1.1 Predicting Which Customers Might be Pregnant

For retailers, early identification of customers who might be pregnant can be profitable. Impending childbirth means there is much to purchase and, as the event approaches, less time for shopping. A retailer that can establish itself early with a pregnant mother stands a good chance of capturing her big-ticket purchases of furniture and clothing. By the time a family is purchasing a crib, stroller, and infant clothing it is too late. The retailer Target decided to use predictive modeling to predict, at an early stage, which customers were pregnant.

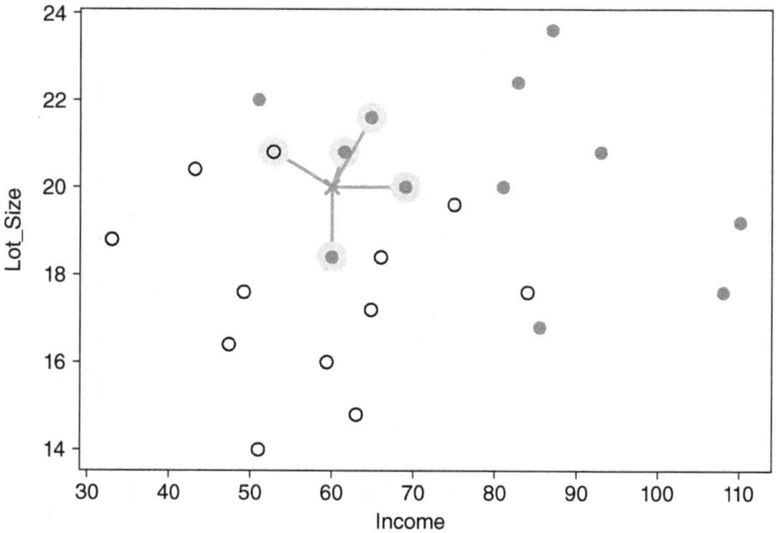

Figure 12.3 Five nearest neighbors ($k = 5$).

The first challenge was to assemble a dataset with known outcomes. For this, Target used its baby shower registry, figuring that all the registrants were, for the most part, pregnant. Then it took a demographically similar set of women from among its general customer records, figuring that they were unlikely to be pregnant. Of course, a few of them would be, but Target still had a dataset with two well-distinguished subgroups—"pregnant" and "not-pregnant."

Next they needed a model to predict, on the basis of past purchases, whether a customer was pregnant or not. Multiple linear regression was not suitable, since the outcome (dependent) variable is not a continuous one, but rather binary. Let's apply k-nearest-neighbors to a hypothetical tiny dataset to illustrate how it works.

12.1.2 Small Hypothetical Example

Consider the data shown in Table 12.1. There is data on six prior customers, two of whom are on the baby shower registry and four who are not. There are six predictor variables (columns) for each customer and one outcome variable. The predictors are 0/1 variables, indicating whether a product (zinc supplements, manganese supplements, cotton swabs) has been bought in the last 10 days, or prior 10–90 day period.

We wish to predict whether the new customer—the row at the bottom—would be on the baby shower registry; this is a proxy for whether the customer is predicted to be pregnant.

Table 12.1 Hypothetical customer purchases.

Cust #	zinc10	zinc90	mag10	mag90	cotton10	cotton90	Registry ?
1	1	1	1	1	0	1	0
2	1	1	1	1	0	1	0
3	1	1	0	1	0	1	0
4	0	1	1	0	1	1	0
5	1	1	0	1	1	0	1
6	0	0	1	0	1	0	1
NEW	1	0	1	1	0	1	?

By eyeballing the data, we can see that the new customer is more like customers 1 and 2 than the other customers—the new customer matches customers 1 and 2 in all predictor variables except one. Customers 1 and 2 are not on the baby shower registry, so we would be inclined to predict that the new customer is *not pregnant*.

With hundreds of predictors and tens of thousands of customers, it is not feasible to eyeball the data, so we need a way to automatically measure how close (alike) customers are. We will use the measure of distance that we first introduced in Section 3.4—sum the squared differences, then take the square root (Euclidean distance).

The Euclidean distance between two vectors w and x is

$$\text{Euclidean Distance} = \sqrt{(w_1 - x_1)^2 + (w_2 - x_2)^2 + \cdots (w_n - x_n)^2}$$

Here are the Euclidean distance calculations for the new customer—compared first to Customer #1 and then to Customer #6:

				Predictor Variables			Outcome
Customer # 1	1	1	1	1	0	1	0
NEW	1	0	1	1	0	1	?
Difference	0	1	0	0	0	0	
Sq difference	0	1	0	0	0	0	
		Sum = 1					
Customer # 6	0	0	1	0	1	0	1
NEW	1	0	1	1	0	1	?
Difference	−1	0	0	−1	1	−1	
Sq difference	1	0	0	1	1	1	
		Sum = 4					

For the new customer and customer #1, the sum of squared differences = 1, so the square root = 1 as well. For customer #6, the sum of squared differences = 4, so the square root = 2. So we conclude that customer #1 (who is not on the baby shower registry) is closer to the new customer.

Note that in this small example, all the data are binary (yes/no) data. Measured data can be used as well. Suppose one we were to add the variable "total spending." For many, if not most, customers, the value of "total spending" will range in the hundreds, if not thousands of dollars. "Total spending" will now dominate the Euclidean distance calculations; the 0's and 1's will be of little consequence. To avoid this, we can normalize (standardize) the data, as discussed in Section 5.4.1. This will put all the variables on the same scale.

The classification procedure is, therefore, as follows:

1) Assemble a "training" data set that includes predictor variables and known outcomes
2) Normalize each variable by subtracting the mean and dividing by the standard deviation (a superfluous step in this simple example where all variables are 0/1, but otherwise needed)
3) For each new customer to be classified,
   a) Calculate and record Euclidean distance to each record in the training data
   b) Note the *k*-closest records
   c) Find which class is prevalent (majority vote) in the *k*-closest records
   d) Classify the new customer as that prevalent class

12.1.3 Setting *k*

As you can see, the results of this method depend on how many neighbors you consult. How do you set *k*? Very small values of *k* respond to highly local information, but may produce unstable classifications (classifications that change depending on the data used to find neighbors). Large values of *k* produce more stable classifications, but may miss local information. Typically you first divide the data into a training set and a holdout set, and use the above procedure to classify the holdout set as "new" data, and find what value of *k* produces the most accurate classifications.

12.1.4 *K*-Nearest-Neighbors and Numerical Outcomes

The above example uses *k*-nearest-neighbors to predict binary outcomes, but the technique can be extended to cover the same sort of measured data we analyzed with regression. Instead of taking a majority vote of class, we would simply take the average of the predicted outcomes among the neighbors.

12.1.5 Explanatory Modeling

K-nearest-neighbors is *not* useful for explanatory modeling, however. We saw in Chapter 11 how we could derive information about linear relationships between predictors and an outcome variable. Other statistical and machine learning models can provide such structural information about such patterns in data: curvilinear relationships, if-then rules, and more. K-nearest-neighbors provides no useful information about "macro-level" relationships—linear or otherwise—between predictor variables and an outcome variable. It only provides the actual predictions.

12.2 Python: Classification

12.2.1 Classification Using `scikit-learn`

For classification, the `scikit-learn` package is the most popular Python package. It implements about 40 different classifier algorithms. Here, we will only look at the k-nearest neighbor algorithm. To be more specific, the method that is implemented by the `sklearn.neighbors.KNeighborsClassifier` class. You will see that fitting a classifier is a lot like fitting a regression model.

We use the *RidingMowers.csv* as an example to demonstrate how to fit a k-nearest neighbor classifier. The outcome variable is `Ownership`, which is a binary variable. The predictors are `Income` and `Lot_Size`. We will use the `KNeighborsClassifier` class to fit the model. However, first we need to normalize the data.

```
import pandas as pd
from sklearn.preprocessing import StandardScaler
from sklearn.neighbors import KNeighborsClassifier

mower_df = pd.read_csv("RidingMowers.csv")
outcome = "Ownership"
predictors = ["Income", "Lot_Size"]

X = mower_df[predictors]
y = mower_df[outcome]

scaler = StandardScaler()
X_normalized = scaler.fit_transform(X)
```

We use the `StandardScaler` class to normalize the predictors. This class is a transformer; it takes data and changes them. Here, this means subtracting the column mean and dividing by the standard deviation. The transformation is first "learned" from the data (*fit*), meaning that the standard deviation and mean are calculated, and then applied to the data by subtracting the mean and dividing by the standard deviation (*transform*). Transformers generally implement these two stages. The *fit_transform* method performs both in one go.

We have now normalized data and can use these to fit the k-nearest neighbor classifier.

```
model = KNeighborsClassifier(n_neighbors=5) ①
model.fit(X_normalized, y) ②
```

① The `KNeighborsClassifier` class implements the k-nearest neighbor classifier. The `n_neighbors` parameter is used to specify the number of neighbors k to use. This type of parameter is known as a tuning parameter or hyperparameter.

② Like in regression models, classification models also use the *fit* method to fit the model to the data.

The `scikit-learn` classifiers also have a *predict* method that we can use to make predictions. Assume we have a new customer with an income of 60 and a lot size of 20. Would this customer buy a riding mower?

```
new_customer = pd.DataFrame({"Income": 60, "Lot_Size": 20},
                            index=["New customer"])
new_customer_normalized = scaler.transform(new_customer)
pred_class = model.predict(new_customer_normalized)
print(f'Class predicted for the new customer: {pred_class[0]}')
```

Output

```
Class predicted for the new customer: Owner
```

In contrast to regression models, classification models have a second type of prediction. The *predict_proba* method returns not a class, but the estimated probability of belonging to each class.

```
pred_class = model.predict_proba(new_customer_normalized)
print(f'Class predicted for the  new customer: {pred_class[0]}')
```

Output

```
Class predicted for the  new customer: [0.2 0.8]
```

The predicted probability for the new customer to be an owner is 0.8 (80%).

12.2.2 Evaluating the Model

We learned that regression models can be evaluated using the R^2 or RMSE statistics. For classification models, the accuracy is a common measure. Accuracy is the proportion of predictions that are correct. In `scikit-learn`, the function *accuracy_score* function calculates the accuracy.

```
from sklearn.metrics import accuracy_score
accuracy = accuracy_score(mower_df[outcome],
   model.predict(X_normalized))
print(f'Accuracy: {accuracy:.2f}')
```

Output
```
Accuracy: 0.83
```

12.2.3 Streamlining Model Fitting Using Pipelines

This classification model required two steps. First, the normalization of the data and then the fitting of the k-nearest neighbor model. Both steps are learned from the data and to predict new data, we need to keep track of both steps. To make this easier, `scikit-learn` has pipelines. Pipelines allow you to link several transformers and estimators together to form a single object.

Let's see how this works.

```
from sklearn.pipeline import Pipeline
model = Pipeline(steps=[
    ('normalize', StandardScaler()),
    ('kNN', KNeighborsClassifier(n_neighbors=5))
])
print(model)
```

Output
```
Pipeline(steps=[('normalize', StandardScaler()),
                ('kNN', KNeighborsClassifier())])
```

The `Pipeline` initializer takes a list of steps and combines them into a pipeline. The steps are defined as tuples, where the first element is the name of the step and the second element is the transformer or estimator. The pipeline has the same methods as the final estimator. When calling `fit`, the data is passed through all transformers and then to the final estimator.

```
model.fit(X, y)
```

As before, the fitted model can predict our new customer. This time we don't have to worry about passing the data through both steps separately—the pipeline handles both the normalization and the model fitting.

```
pred_class = model.predict(new_customer)
print(f'Class predicted for the  new customer: {pred_class[0]}')
```

Output
```
Class predicted for the  new customer: Owner
```

Next calculate the accuracy of the model.

```
accuracy = accuracy_score(mower_df[outcome], model.predict(X))
print(f'Accuracy: {accuracy:.2f}')
```

Output

```
Accuracy: 0.83
```

There is a lot more to pipelines and `scikit-learn` in general that we won't cover here. For example, there are many more classifiers, transformers, and tools for model evaluation. The `scikit-learn` documentation is a great resource to learn more about these topics. The book *Data Mining for Business Analytics* by Shmueli et al. (2019, Wiley, 1st edition)[1] is also a great resource to learn more about `scikit-learn` and other Python packages for machine learning and data mining.

Exercises

12.1 If you want your predictions using k-NN to be quite sensitive to the information contained in the most similar records, would you choose k to be relatively low or relatively high?

12.2 If you want your predictions using k-NN to be quite stable and not dependent on random elements in the chosen sample, would you choose k to be relatively low or relatively high?

12.3 This exercise with a tiny dataset illustrates the calculation of Euclidean distance and the creation of binary dummies. The online education company Statistics.com segments its customers and prospects (noncustomers) into three main professional categories: software engineers (EN), data scientists (DS), and other. They are also categorized by whether they are located in the United States, and whether they first came to Statistics.com after being referred by another customer. Consider Table 12.2, showing the customers and prospects and the additional information about whether they have purchased a course or not.
Consider now the following new prospect, for whom you want to predict whether they will become a customer:
Prospect 1: EN, located in the United States, referred by another customer

1 A revised second edition will be published in 2024 under the name *Machine Learning for Business Analytics: Concepts, Techniques and Applications in Python* (2024, Wiley, 2nd edition).

Table 12.2 Online course customers.

Customer #	Profession	US?	Referred?	Bought Course?
1	DS	0	1	0
2	DS	1	0	1
3	EN	0	0	1
4	DS	1	1	0
5	Other	1	0	1

a) Just looking at the five customers and prospects, and considering the data on profession, location, and referral, which is most similar (closest) to the new prospect?
b) Convert the Profession variable into 3 binary dummies (see Section 11.6.4), and calculate Euclidean distance between the new prospect and each of the five known customers and prospects. Which is closest? (*Note*: While it is typical to normalize data for k-NN, this is not needed here since the variables are binary and all on the same scale.)

12.4 Personal Loan Acceptance: Universal Bank is a relatively young bank growing rapidly in terms of overall customer acquisition. The majority of these customers are liability customers (depositors) with varying sizes of relationship with the bank. The customer base of asset customers (borrowers) is quite small, and the bank is interested in expanding this base rapidly to bring in more loan business. In particular, it wants to explore ways of converting its liability customers to personal loan customers (while retaining them as depositors).
A campaign that the bank ran last year for liability customers showed a healthy conversion rate of over 9% success. This has encouraged the retail marketing department to devise smarter campaigns with better target marketing. The goal is to use k-NN to predict whether a new customer will accept a loan offer.
This will serve as the basis for the design of a new campaign.
The file *universal-bank.csv* contains data on 5000 customers. The data include customer demographic information (age, income, etc.), the customer's relationship with the bank (mortgage, securities account, etc.), and the customer response to the last personal loan campaign (Personal Loan). Among these 5000 customers, only 480 (= 9.6%) accepted the personal loan that was offered to them in the earlier campaign.

Transform categorical predictors with more than two categories into dummy variables and normalize the data.

a) Consider the following customer:
   Age = 40, Experience = 10, Income = 84, Family = 2, CCAvg = 2, Education_1 = 0, Education_2 = 1, Education_3 = 0, Mortgage = 0, Securities Account = 0, CD Account = 0, Online = 1, and Credit Card = 1. Perform a k-NN classification with all predictors except ID and ZIP code using $k = 1$. Specify the *success* class as 1 (loan acceptance), and use the default cutoff value of 0.5 (i.e. straight majority vote). How would this customer be classified?
b) Repeat this classification using $k = 5$ and $k = 9$ and report the results. What are the considerations in setting k?

Index

Statistics for Data Science and Analytics, First Edition. Peter C. Bruce, Peter Gedeck, and Janet Dobbins.

Companion website: www.wiley.com/go/Wiley_Statistics_for_Data

www.ingramcontent.com/pod-product-compliance
Lightning Source LLC
Chambersburg PA
CBHW071057140426
43066CB00002BA/4
* 9 7 8 1 3 9 4 2 5 3 8 0 7 *